Design Optimization Using Artificial Intelligence

Design Optimization Using Artificial Intelligence serves as an insightful resource for understanding the transformative role of Artificial Intelligence (AI) in optimizing design processes across a variety of fields. This book explores foundational concepts, advanced methodologies, and real-world applications, offering a comprehensive guide to leveraging AI for innovative design solutions.

Covering a broad spectrum of topics, the book examines algorithmic advancements, machine learning techniques, and optimization strategies that enhance design efficiency and accuracy. It delves into AI's applications in sustainable energy systems, biomedical innovations, image processing, cybersecurity, and other critical domains, showcasing its ability to solve complex challenges with precision and creativity.

The content bridges theoretical foundations with practical implementations, incorporating case studies that highlight successful applications of AI-driven design optimization. Readers will gain valuable insights into state-of-the-art tools, programming techniques, and frameworks that drive advancements in intelligent design systems.

Features:

- Explores AI-driven methodologies for enhancing design accuracy and efficiency
- Discusses advanced machine learning techniques and optimization algorithms
- Highlights AI's applications in diverse fields such as energy systems, healthcare, image processing, and cybersecurity
- Includes in-depth case studies to illustrate real-world implementations of AI in design optimization
- Investigates future trends and the potential of AI to transform traditional design processes

Targeted at researchers, engineers, and students in the fields of artificial intelligence, computational science, and engineering, this book is an essential reference for anyone seeking to understand and utilize AI in the realm of design optimization. It provides tools and knowledge necessary to harness AI's power for innovative, efficient, and intelligent design solutions.

Artificial Intelligence and Machine Learning for Intelligent Engineering Systems

Series editors: Mukesh Kumar Awasthi, Ravi Tomar, and Ashwani Kumar

The book series editors are inviting edited, reference, and textbook proposals for consideration in the book series. This book series highlights into the profound impact of AI and ML across various engineering disciplines, focusing on mechanical, materials, electrical, computer, and electronics engineering. Each volume explores deep into specific applications tailored to each field. For instance, the series explores how AI and ML algorithms can predict equipment failures and optimize maintenance schedules, leading to significant benefits for both mechanical and electrical systems. In materials science, the series investigates how machine learning can be harnessed to discover innovative materials with targeted properties, accelerating the pace of discovery. Similarly, the application of AI for intelligent control of complex systems in electrical and mechanical engineering is addressed, paving the way for enhanced performance and efficiency. Additionally, the series explores how machine learning can be employed to optimize designs across mechanical, electrical, and computer systems, taking into account factors like functionality, ease of manufacturing, and energy consumption. Furthermore, the series sheds light on the utilization of machine learning algorithms to analyze sensor data from diverse engineering systems, enabling real-time diagnostics and proactive fault detection. Finally, the integration of AI and ML into robotics is explored, highlighting its potential for enhanced decision-making, path planning, and fostering a more collaborative human-machine interface. This comprehensive series serves as an invaluable resource for engineers, researchers, and students across various engineering disciplines. It provides a thorough overview of the latest advancements in AI and ML, along with practical applications for designing and developing intelligent engineering systems that will shape the future.

Design Optimization Using Artificial Intelligence
Edited by Satya Ranjan Mishra, Apul Narayan Dev, Alok Kumar Pandey, and Mukesh Kumar Awasthi

For more information about this series, please visit: www.routledge.com/Artificial-Intelligence-and-Machine-Learning-for-Intelligent-Engineering/book-series/AIMLIES

Design Optimization Using Artificial Intelligence

Edited by
Satya Ranjan Mishra
Apul Narayan Dev
Alok Kumar Pandey
Mukesh Kumar Awasthi

CRC Press
Taylor & Francis Group
Boca Raton London New York

CRC Press is an imprint of the
Taylor & Francis Group, an **informa** business

Designed cover image: Shutterstock

First edition published 2025
by CRC Press
2385 NW Executive Center Drive, Suite 320, Boca Raton FL 33431

and by CRC Press
4 Park Square, Milton Park, Abingdon, Oxon, OX14 4RN

CRC Press is an imprint of Taylor & Francis Group, LLC

ISBN: 978-1-032-96494-2 (hbk)
ISBN: 978-1-032-96495-9 (pbk)
ISBN: 978-1-003-58971-6 (ebk)

DOI: 10.1201/9781003589716

Typeset in Sabon
by SPi Technologies India Pvt Ltd (Straive)

Contents

About the editors

Satya Ranjan Mishra is currently working as a Professor in the Department of Mathematics at Siksha O Anusandhan Deemed to be University. He has 19 years of teaching experience at both the postgraduate and undergraduate levels. He has completed his doctoral studies at Siksha O Anusandhan in 2013 and has been actively engaged in research since then. His areas of interest include Heat Transfer, Magnetohydrodynamics, Porous Media, etc., within the broader field of Fluid Dynamics. He has published nearly 250 papers in international journals of repute, which are either SCI or Scopus indexed and can be found in various databases such as Scopus, ResearchGate, Google Scholar, etc. Due to the significant citation of his work, he was recognized as one of the Top 2% World Scientists by Stanford University, USA, for four consecutive years (2020–2023). He has also guided 13 research scholars, with six more students currently working under his guidance. Dr. Mishra has delivered several lectures at various conferences and faculty development programs, including one organized by Rajasthan Technical University, Rajasthan, in 2019 on the topic "Introduction to MATLAB," an FDP organized by Poornima Group of Institutions, Rajasthan, on the topic "Numerical Solutions using the Scientific Tool MAPLE" in 2020 (online), and a Summer Instructional School organized by NIT Arunachal Pradesh in 2020 on the topic "Linear Algebra." He is the series editor for a book series at Springer Nature and has also edited a book under it in the year 2020. He is also an editor/associate editor for several reputed international journals. Dr. Mishra has authored a book titled *Learning Numerical Methods using MATLAB* and is expected to be released soon.

Apul Narayan Dev is currently working as an Professor, Department of Mathematics, Siksha 'O' Anusandhan Deemed to be University, Bhubaneswar. He has 14 years of teaching experience at both the postgraduate and undergraduate levels. He has completed his MPhil and PhD from Gauhati University in 2011 and 2016, respectively. His areas of interest include the basic study of theoretical plasmas, degenerate and non-degenerate plasmas, mathematical methods, and fluid flow, with a broad focus on plasmas and fluid dynamics. He has published nearly 48 papers in reputable international journals, all of which are either SCI or Scopus indexed, and can be accessed through databases like Scopus, ResearchGate, Google Scholar, etc. He also published a book titled *Shock Wave Phenomena in Dusty Plasma* which is available on Amazon. Dr. Dev has guided one research scholar, with four more students working under his guidance. Additionally, he has organized Three international conferences (AMSE-2019, AMSE-2022, AMSE-2024) and two international webinars (Maple-2020, Mathematica-2020), and edited a Scopus-indexed conference proceeding published by a reputed publisher in 2022 as the leading editor.

Alok Kumar Pandey is an Assistant Professor in the Department of Mathematics at Graphic Era (Deemed to be University), Dehradun, Uttarakhand, India. He obtained his PhD from Govind Ballabh Pant University of Agriculture and Technology, Pantnagar, Uttarakhand, India. His research area is Computational Fluid Dynamics. He has published over 60 research articles in international journals and was included in the World's Top 2% Scientists 2023 list by Stanford University. In 2018, he was awarded the Publons Peer Review Award. He is a certified reviewer for more than 100 international journals and currently serves as an editor for *Open Physics* and *Journal of Engineering Researcher and Lecturer*, associate editor for the *Journal of Advanced Research in Numerical Heat Transfer* and *Journal of Advanced Research in Micro and Nano Engineering*, and a member of the editorial board for journals like *Teknomekanik* and *SCIREA Journal of Mathematics*. His scientific metrics, according to Google Scholar, show an h-index of 30, 2,126 citations, and an i10-index of 43.

Mukesh Kumar Awasthi completed his PhD, and his thesis is titled "Viscous Correction for the Potential Flow Analysis of Capillary and Kelvin-Helmholtz Instability." He is currently an Assistant Professor in the Department of Mathematics at Babasaheb Bhimrao Ambedkar University, Lucknow. Dr. Awasthi specializes in the mathematical modeling of flow problems. He has taught courses in Fluid Mechanics, Discrete Mathematics, Partial Differential Equations, Abstract Algebra, Mathematical Methods, and Measure Theory to postgraduate students. He has extensive knowledge of mathematical modeling in fluid dynamics and can solve flow problems analytically and numerically. His areas of expertise include viscous potential flow, electro-hydrodynamics, magneto-hydrodynamics, heat, and mass transfer. Dr. Awasthi has excellent communication skills and leadership qualities and is self-motivated, responding to suggestions in a constructive manner. He qualified the National Eligibility Test (NET) conducted by the Council of Scientific and Industrial Research (CSIR) in 2008, earning a Junior Research Fellowship (JRF) and Senior Research Fellowship (SRF). He has published over 125 research publications, including journal articles, books, book chapters, and conference papers, in national and international journals and conferences. He has also published 10 books and is a series editor for *Artificial Intelligence and Machine Learning for Intelligent Engineering Systems*, published by CRC Press (Taylor & Francis, USA). He has attended numerous symposia, workshops, and conferences in mathematics and fluid mechanics. Dr. Awasthi received "Research Awards" consecutively from 2013 to 2016 from the University of Petroleum and Energy Studies, Dehradun, India, and has also received a start-up research fund from the UGC, New Delhi, for his project titled "Nonlinear Study of the Interface in Multilayer Fluid Systems." He has been recognized as one of the top 2% influential researchers globally by Stanford University for 2022 and 2023. His ORCID is 0000-0002-6706-5226, and his Google Scholar and ResearchGate profiles can be accessed at Google Scholar and ResearchGate, respectively.

Contributors

Abdulhalim Musa Abubakar
Department of Chemical Engineering, Faculty of Engineering
Modibbo Adama University
Adamawa State, Nigeria

Mukesh Kumar Awasthi
Department of Mathematics
Babasaheb Bhimarao Ambedkar University
Lucknow, India

Swadhin Kumar Barisal
Department of Computer Application
Siksha 'O' Anusandhan (Deemed to be University)
Bhubaneswar, India

Muhammad Tayyab Bilal
Computer Science Department
University of Agriculture
Faisalabad Punjab, Pakistan

Rajen Bose
Department of CSE
GIET University
Gunupur, India

Chaahat
Department of Computer Science and Engineering
Chitkara University
Punjab, India

Chandreyee Chakroborty
St. Thomas' College of Engineering and Technology
Kolkata, WestBengal

Subhashis Chatterjee
Department of Mathematics & Computing
Indian Institute of Technology (Indian School of Mines)
Dhanbad, India

Ayes Chinmay
Department of Computer Science and Engineering
ITER, Siksha 'O' Anusandhan (Deemed to be University)
Bhubaneswar, India

Avisikta Mandal Das
Department of Computer Science and Engineering
ITER, Siksha 'O' Anusandhan (Deemed to be University)
Bhubaneswar, India

Niva Das
Department of Electronics and Telecommunication Engineering
ITER, Siksha 'O' Anusandhan (Deemed to be University)
Bhubaneswar, India

Soubhagya Das
Department of CSE
GIET University
Gunupur, India

Subhadarshini Das
Department of ECE
ITER, Siksha 'O' Anusandhan (Deemed to be University)
Bhubaneswar, India

Umesh Kumar Das
Department of Computer Science & Engineering
Siksha O Anusandhan (Deemed to be University)
Bhubaneswar, India

Rajashree Dash
Department of Computer Science & Engineering
Siksha O Anusandhan (Deemed to be University)
Bhubaneswar, India

Saurav Datta
ITER, Siksha 'O' Anusandhan (Deemed to be University)
Bhubaneswar, India

Apul N. Dev
Department of Mathematics
Siksha 'O' Anusandhan (Deemed to be University)
Bhubaneswar, India

Tanka Nath Dhamala
Central Department of Mathematics
Tribhuvan University
Kathmandu, Nepal

Issam Ferhoune
Department of Process Engineering
Faculty of Science and Applied Sciences, University of Oum El Bouaghi
'Larbi Ben M'hidi', Algeria

Yatika Gori
Department of Mechanical Engineering
Graphic Era deemed to be University Dehradun
India

Lambodar Jena
Department of Computer Science and Engineering
Siksha 'O' Anusandhan (Deemed to be University)
Bhubaneswar, India

Mohuya Byabartta Kar
Heritage Institute of Technology
Kolkata, India

Narendra Kumar Kamila
Department of Computer Science and Engineering (DS)
GITA Autonomous College
Bhubaneswar, India

Durga Prasad Khanal
Saraswati Multiple Campus, Tribhuvan University
Kathmandu, Nepal

Ashwani Kumar
Technical Education Department Uttar Pradesh
Kanpur, India

Amiya Maity
Department of Computer Science & Engineering
Siksha O Anusandhan (Deemed to be University)
Bhubaneswar, India

Biplab Mandal
Meghnad Saha Institute of Technology
Kolkata, India

Bibhuprasad Mohanty
Department of ECE
ITER, Siksha 'O' Anusandhan (Deemed to be University)
Bhubaneswar, India

Mihir Narayan Mohanty
Department of ECE
ITER, Siksha 'O' Anusandhan (Deemed to be University)
Bhubaneswar, India

Smita Mohanty
Department of Computer Science & Engineering
Siksha O Anusandhan (Deemed to be University)
Bhubaneswar, India

Amar Jyoti Nayak
Computer Science and Engineering
ITER, SOAU
Bhubaneswar, India

Pratyush Nayak
Computer Science and Engineering
ITER, SOAU
Bhubaneswar, India

Pratyush Ranjan Nayak
ITER, Siksha 'O' Anusandhan (Deemed to be University)
Bhubaneswar, India

Soumen Nayak
Department of Computer Science and Engineering
Siksha 'O' Anusandhan (Deemed to be University)
Bhubaneswar, India

Simisani Ndaba
Department of Computer Science
University of Botswana
Gaborone, Botswana

Hemanta Kumar Palo
Department of Electronics and Communication Engineering
Institute of Technical Education and Research
Siksha O Anusandhan (Deemed to be University)
Bhubaneswar, Odisha, India

Susmita Panda
Computer Science and Engineering
ITER, SOAU
Bhubaneswar, India

Ankita Pandey
Department of Mathematics
GLA University
Mathura, India

Anmol Pattanaik
Department of Computer Science and Engineering
ITER, Siksha 'O' Anusandhan (Deemed to be University)
Bhubaneswar, India

Niharika Pattanaik
Department of Electronics and Communication Engineering
Institute of Technical Education and Research
Siksha O Anusandhan (Deemed to be University)
Bhubaneswar, Odisha, India

Neha Singh Raghuvanshi
ABES Engineering College
Ghaziabad, India

Unnayan Raj
ITER, Siksha 'O' Anusandhan (Deemed to be University)
Bhubaneswar, India

Bidushkumar Sahoo
Department of CSE
GIET University
Gunupur, India

Lokanath Sarangi
Department of Electronics and Communication Engineering
Orissa Engineering College
Bhubaneswar, Odisha, India

Zakiyyu Muhammad Sarkinbaka
Department of Chemical Engineering
Federal University Wukari (FUW)
Wukari, Nigeria

Sadisha Sasneha
Computer Science and Engineering
ITER, SOAU
Bhubaneswar, India

Namita Shaw
Department of Computer Science and Engineering
ITER, Siksha 'O' Anusandhan (Deemed to be University)
Bhubaneswar, India

Manas Kumar Swain
Department of Computer Science and Information Technology
Siksha 'O' Anusandhan (Deemed to be University)
Bhubaneswar, India

Shreya Swarnaker
Department of Mathematics & Computing
Indian Institute of Technology (Indian School of Mines)
Dhanbad, India

Debashish Tiwary
Department of Computer Science and Engineering
ITER, Siksha 'O' Anusandhan (Deemed to be University)
Bhubaneswar, India

Binodini Tripathy
Department of Electronics and Telecommunication Engineering
KP, KIIT Deemed to be University
Bhubaneswar, India

Hemant Yadav
Department of Mechanical Engineering
GBPUAT Pantnagar (U.S. Nagar)
India

Shailesh Yadav
Department of Mechanical Engineering
GBPUAT Pantnagar (U.S. Nagar)
India

Aim and scope

The book *Design Optimization Using Artificial Intelligence* aims to explore the integration of artificial intelligence (AI) techniques with traditional design optimization methods across various engineering fields. The primary goal is to present AI as a powerful tool to improve design processes by automating decision-making, enhancing performance, and optimizing resource allocation. By combining cutting-edge AI techniques with classical design approaches, this book serves as a comprehensive resource for researchers, engineers, and practitioners interested in applying AI to complex optimization problems in design.

Chapter 1 introduces recent developments in AI-powered mechanical design and optimization. This chapter sets the foundation by highlighting how AI has evolved to address mechanical design challenges, focusing on how these advancements have revolutionized optimization techniques and broadened the scope of what can be achieved in design processes. In **Chapter 2**, the book covers the fundamentals of machine learning and its direct application to design optimization. This chapter lays the groundwork for understanding how machine learning algorithms, such as neural networks, decision trees, and support vector machines, are applied to optimize design parameters, offering insights into the key principles and techniques involved. **Chapter 3** provides detailed case studies where machine learning techniques have been successfully applied in design optimization. These case studies exemplify the practical application of the concepts discussed in previous chapters, illustrating real-world challenges and how machine learning offers effective solutions in various industries, including automotive, aerospace, and manufacturing. **Chapter 4** delves into the role of artificial intelligence in manufacturing and production. By exploring how AI optimizes production processes, improves quality control, and enhances supply chain management, this chapter highlights AI's significance in making manufacturing more efficient, sustainable, and cost-effective. In **Chapter 5**, the book addresses how algorithmic modifications enhance text classification in machine learning. A comparative study of AI-driven optimization techniques is presented, offering readers a deep understanding of how text classification can be improved using novel AI strategies, providing a unique perspective on AI applications in data-driven optimization tasks.

Chapter 6 focuses on enhancing the performance of deep neural networks by utilizing ANOVA-based feature selection for drought prediction. This chapter demonstrates how feature selection techniques can be used to fine-tune AI models, ensuring better prediction accuracy, and explores how AI models can contribute to solving global challenges like drought prediction. **Chapter 7** covers protein-protein interaction prediction using extended natural vector representations of proteins, comparing it with natural vector methods. The chapter explores how AI and machine learning can be used in bioinformatics for understanding protein interactions, with a focus on improving accuracy and predicting complex biological relationships. **Chapter 8** introduces an optimized Chebyshev Polynomial Neural Network (CHPNN) model for COVID-19 case prediction. It highlights how AI-driven models can predict the course of the pandemic, showcasing the power of AI in handling public health crises by forecasting key metrics like confirmed cases, recoveries, and fatalities. In **Chapter 9**, the book discusses an optimized filter design using fractional anisotropic diffusion techniques for image denoising. This chapter focuses on how AI techniques like image processing can be optimized to enhance image quality in fields like medical imaging, where accurate data interpretation is critical. **Chapter 10** addresses the role of enhanced visual cryptography algorithms in cybersecurity. The chapter introduces novel cryptographic techniques that utilize AI to protect sensitive data, ensuring secure communications and privacy in digital systems, a critical aspect in today's increasingly connected world.

Chapter 11 explores the use of i-vector-based radial basis function networks for effective classification of brain tumors. This chapter delves into how AI-driven models can automate and improve the accuracy of medical diagnoses, particularly in identifying and classifying brain tumors from medical imaging data. In **Chapter 12**, deep transfer learning with AlexNet is employed to enhance breast cancer detection. This chapter showcases how deep learning algorithms can be used to improve medical diagnostics, particularly in detecting abnormalities in mammography images, thus contributing to early diagnosis and better patient outcomes. **Chapter 13** focuses on the identification of brain tumors using DenseNet-201 deep learning models. By examining how advanced deep learning models can be used to automatically detect brain tumors with high accuracy, this chapter adds value to medical applications of AI, emphasizing the potential of these models in clinical settings. In **Chapter 14**, IoT-powered video surveillance systems are enhanced using WiMAX technology for improved security. This chapter highlights how the combination of AI and IoT can provide effective surveillance systems for enhancing security, especially in critical infrastructure and smart cities. **Chapter 15** discusses the implementation of automatic gender detection using AI, particularly through speech and audio analysis. The chapter explains how machine learning classifiers can be trained to accurately detect gender, showing how AI can be applied in areas like biometric security and human-computer interaction.

Chapter 16 presents a solution for optimal dynamic contraflow with intermediate storage using anti-parallel path decomposition. This chapter delves into optimization problems related to traffic flow, demonstrating how AI algorithms can improve transportation systems by optimizing flow management and storage allocation in dynamic environments. Finally, **Chapter 17** explores heart disease prognosis using machine learning classifiers with hyperparameter optimization. It illustrates how machine learning can be applied to healthcare, specifically in predicting heart disease risk, showcasing the ability of AI to improve decision-making and diagnostic accuracy in critical medical conditions.

Through its well-rounded chapters, the book offers a broad and comprehensive perspective on how AI can be integrated into design optimization processes. It provides both theoretical foundations and practical applications, making it a valuable resource for anyone looking to leverage AI in solving complex optimization problems across various fields.

Satya Ranjan Mishra
Apul Narayan Dev
Alok Kumar Pandey
Mukesh Kumar Awasthi

Preface

The field of design optimization has seen tremendous advancements with the rise of Artificial Intelligence (AI). The integration of AI into optimization processes has revolutionized industries across the globe, enhancing efficiency and performance in various engineering domains. The book *Design Optimization Using Artificial Intelligence* explores the intersection of AI and optimization techniques, offering a comprehensive approach to understanding and applying AI in design challenges. This book is a valuable resource for students, researchers, and professionals looking to enhance their knowledge and skills in this rapidly evolving field.

Chapter 1 lays the foundation by examining recent developments in AI-powered mechanical design and optimization. It introduces the core concepts and discusses how AI techniques are being applied to mechanical design processes to achieve optimal solutions in real-world applications. Chapter 2 focuses on the fundamentals of machine learning and its crucial role in design optimization. It covers the core principles of machine learning algorithms, highlighting their potential for optimizing design processes and solving complex engineering problems. In Chapter 3, we present a series of case studies that demonstrate the practical applications of machine learning for design optimization. These case studies offer real-world examples of how AI-driven techniques have been successfully applied across various industries to improve design outcomes. Chapter 4 dives into the application of AI in manufacturing and production. It emphasizes the role of AI in optimizing manufacturing processes, reducing costs, and enhancing production efficiency, and explores its potential for transforming the industrial landscape. Chapter 5 examines AI-driven optimization techniques in text classification, offering a comparative study of algorithmic modifications to enhance classification performance. It provides insights into how these techniques can be applied to optimization problems beyond the traditional design applications.

Chapter 6 focuses on enhancing the performance of deep neural networks through feature selection for drought prediction. It discusses the significance of AI in predicting environmental changes and how optimizing deep learning models can improve prediction accuracy. Chapter 7 presents the application of AI in protein-protein interaction prediction, utilizing an extended natural vector representation of proteins. The chapter compares different methods and highlights AI's role in biological data analysis, which is essential for drug discovery and disease prevention. In Chapter 8, we explore the use of an optimized CHPNN model for COVID-19 case prediction. This chapter illustrates how AI techniques are leveraged in the healthcare domain to predict the spread of infectious diseases and improve public health response strategies. Chapter 9 discusses the design of optimized filters using fractional anisotropic diffusion techniques for image denoising in AI environments. It covers the methods used to enhance image quality and provides practical applications of AI in digital image processing. Chapter 10 explores the role of enhanced visual cryptography algorithms in cybersecurity, addressing how AI techniques can secure sensitive information through advanced encryption methods and contribute to a safer digital environment.

Chapter 11 demonstrates effective brain tumor classification using an i-vector-based radial basis function network. It highlights how AI can assist in medical diagnostics, particularly in the early detection of serious health conditions like brain tumors. In Chapter 12, the focus shifts to enhancing breast cancer detection using deep transfer learning with AlexNet. The chapter discusses the potential of AI in improving the accuracy of cancer detection, ultimately leading to better healthcare outcomes. Chapter 13 introduces the DenseNet-201 deep learning model for brain tumor identification. It presents a state-of-the-art AI model that offers significant improvements in diagnosing and classifying medical images for more accurate results. Chapter 14 addresses the integration of IoT and WiMAX technology in video surveillance systems. It illustrates how AI and IoT can work together to improve security and monitoring systems, providing real-time analytics and advanced security features. In Chapter 15, automatic gender detection using AI techniques is explored, focusing on speech recognition and machine learning classifiers. The chapter demonstrates how AI can be applied to biometric systems for more efficient and accurate gender classification.

Chapter 16 presents an optimal dynamic contraflow model with intermediate storage using anti-parallel path decomposition. The chapter explains how AI can be used to optimize transportation networks and improve flow efficiency in complex systems. Finally, Chapter 17 discusses the use of machine learning classifiers with hyperparameter optimization for heart disease prognosis. It emphasizes AI's role in healthcare, particularly in predicting heart disease risks and supporting medical decision-making processes.

This book offers a unique blend of theoretical insights and practical applications, making it an essential guide for anyone interested in the intersection of AI and design optimization. It is our hope that the readers will find this work valuable and inspiring for their own endeavors in this rapidly growing field.

Satya Ranjan Mishra
Apul Narayan Dev
Alok Kumar Pandey
Mukesh Kumar Awasthi

Acknowledgments

The completion of this book, *Design Optimization Using Artificial Intelligence*, would not have been possible without the invaluable contributions of numerous individuals and organizations.

First and foremost, we would like to express our deepest gratitude to the contributing authors. Their expertise, dedication, and in-depth research have enriched this book with innovative approaches to design optimization using AI. Their commitment to pushing the boundaries of knowledge has been crucial to the success of this project.

We are also profoundly thankful to our academic and professional colleagues for their insightful feedback and continuous support throughout the development of this book. Their constructive reviews and suggestions have played a significant role in enhancing the quality of the work. We are especially appreciative of the peer reviewers, whose thorough evaluations helped shape the chapters into a more refined and comprehensive resource.

Our sincere appreciation extends to our institutions for their unwavering support, encouragement, and provision of the necessary resources. The collaborative and nurturing environment within these institutions has been instrumental in the completion of this project.

We also wish to thank the publishers for their confidence in this work and for their expert guidance throughout the publication process. Their professional assistance, from editorial oversight to production management, has been essential in transforming this book into its final form.

Finally, we would like to express our heartfelt thanks to our families and friends for their constant support and encouragement during the writing and research phases. Their understanding and patience, particularly during the long hours spent on this project, have been an incredible source of strength.

This book is the result of collective efforts, and we are truly grateful to everyone who has contributed to its creation. We hope that *Design Optimization Using Artificial Intelligence* serves as a valuable resource for researchers, students, and professionals, advancing the field of artificial intelligence and its application to design optimization.

Editors
Satya Ranjan Mishra
Apul Narayan Dev
Alok Kumar Pandey
Mukesh Kumar Awasthi

Chapter 1

Recent developments in AI-powered mechanical design and optimization

Ashwani Kumar, Mukesh Kumar Awasthi, S. P. Singh, and Neha Singh Raghuvanshi

1.1 INTRODUCTION

Mechanical design, a cornerstone of mechanical engineering, involves the intricate process of transforming resources and energy into functional mechanical shapes or processes. This transformative endeavor aims to create machines and equipment that cater to human needs and demands. The design process encompasses both the creation of entirely new machines and the optimization of existing ones. This optimization involves the application of innovative ideas and techniques to enhance performance, efficiency, and reliability. To support this complex process, a wide range of powerful software solutions, such as CAx systems, have emerged. These systems empower engineers to simulate and analyze various stages of a machine's lifecycle, from initial concept to final production, within a virtual environment [1, 2].

The machine designer plays a pivotal role in transforming abstract ideas into tangible realities. They are responsible for ensuring the device's functionality, desired features, and cost-effectiveness. However, the increasing complexity of technical challenges often hinders the design process. Time constraints, budgetary limitations, and knowledge gaps can lead to design errors. To address these challenges and streamline the design process, the integration of automation, artificial intelligence, and machine learning becomes imperative. These technologies significantly reduce development time, minimize costs, and drastically lower the error rate [3]. By automating information processing through algorithms, the reliance on human intervention is diminished, leading to more efficient and accurate design outcomes.

This study offers a comprehensive exploration of machine learning and deep learning techniques, providing specific recommendations for their application in the design of machine components and nodes. Machine learning, a cornerstone of artificial intelligence, has the potential to revolutionize mechanical design and optimization. By identifying opportunities for its widespread application, this research aims to contribute to societal advancement and the future of mechanical engineering. Machine learning and deep learning are already integral to various aspects of daily life, with ongoing research dedicated to enhancing their performance and capabilities. For

DOI: 10.1201/9781003589716-1

instance, researchers have developed cluster-based multidimensional approaches to detect attacks on connected vehicles, employing neural network-based anomaly detection systems [4, 5]. Furthermore, machine learning and artificial intelligence have made significant strides in the field of diagnostics, as exemplified by a 2020 review that proposed a comprehensive diagnostic procedure for machine health detection [6].

1.2 DEVELOPMENTS RELATED TO AI, ML AND DL

1.2.1 Artificial intelligence (AI)

Artificial intelligence, a field that has captivated researchers since the 1950s, is experiencing a surge in interest due to its potential to analyze and leverage vast amounts of data. Advancements in machine learning (ML), coupled with increased data storage and computational power, have significantly contributed to this growth. AI's influence extends across various domains, from self-driving cars and chess-playing machines to theorem provers and artistic creations. As a dynamic and multifaceted discipline within computer science, AI holds immense promise for future innovations. One of the primary goals of AI is to develop intelligent machines capable of mimicking human-like behaviors, including learning, reasoning, and problem-solving.

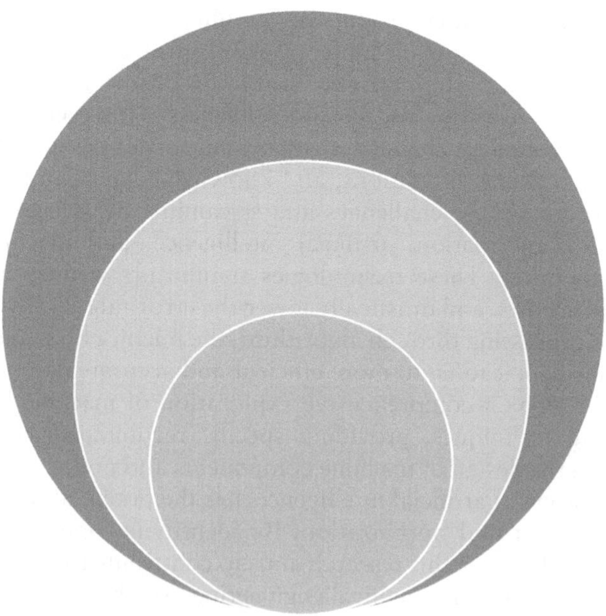

Figure 1.1 Relations of AI, ML and DL.

Table 1.1 shows the advantages, disadvantages, limitations, and applications of Artificial Intelligence (AI) [7–9].

1.2.2 Machine learning (ML)

Machine learning (ML), a subset of artificial intelligence, leverages data and algorithms to mimic human learning processes and progressively enhance accuracy. As a pivotal component of data science, ML empowers organizations to extract valuable insights from large datasets. Through the application of classification and prediction algorithms, data mining initiatives enable informed decision-making, potentially driving significant growth. The exponential growth of big data underscores the increasing demand for skilled data scientists who can effectively identify critical business challenges and deliver actionable solutions [15–18]. Table 1.2 shows the advantages, disadvantages, limitations, and applications of ML.

Table 1.1 Features of Artificial Intelligence (AI) [10, 11]

Feature	*Artificial Intelligence (AI)*
Advantages	Artificial intelligence (AI) offers several potential benefits, including automating repetitive tasks to increase efficiency and productivity, analyzing vast datasets to extract valuable insights for informed decision-making, solving complex problems beyond human capabilities, and personalizing experiences to deliver tailored products and services.
Disadvantages	The development and implementation of AI systems can be financially demanding, while also raising ethical concerns regarding bias, privacy, and potential job displacement. Furthermore, AI-driven automation may lead to job losses in specific industries.
Limitations	AI models require high-quality data to function effectively and they often lack common sense and may struggle with real-world situations. Some AI models, especially deep learning models, can be difficult to interpret and understand.
Applications [12–14]	Artificial intelligence (AI) has permeated various industries, revolutionizing processes and driving innovation. In healthcare, AI is employed for medical diagnosis, drug discovery, and personalized medicine. In finance, AI facilitates fraud detection, algorithmic trading, and risk assessment. Autonomous vehicles, powered by AI, are transforming transportation, while AI-powered chatbots and virtual assistants enhance customer service. In manufacturing, AI optimizes quality control, predictive maintenance, and supply chain management.

Table 1.2 Features of Machine Learning (ML) [19–23]

Feature	*Machine Learning (ML)*
Advantages	Machine Learning (ML) offers a suite of techniques to automate repetitive tasks, thereby improving efficiency and productivity. By analyzing vast datasets, ML algorithms can uncover hidden patterns and trends, enabling data-driven decision-making. Additionally, ML models can predict future outcomes based on historical data and personalize experiences to individual preferences.
Disadvantages	• ML models rely on large amounts of high-quality data to train effectively. • ML models can be complex and difficult to interpret. • Computational cost: Training and deploying ML models can be computationally expensive.
Limitations	Machine Learning (ML) models often require extensive data to achieve optimal performance. However, ML models can be susceptible to biases present in the training data, potentially leading to discriminatory or inaccurate outcomes. Additionally, some ML models, particularly deep learning models, can be complex and difficult to interpret, making it challenging to understand their decision-making processes and identify potential sources of error.
Applications	• ML is used for image and speech recognition in applications like facial recognition and voice assistants. • ML is used for tasks like machine translation, text summarization, and sentiment analysis. • ML is used to recommend products, movies, and other content to users. • ML is used to detect fraudulent transactions and activities. • ML is used to analyze medical images and patient data to aid in diagnosis.

Table 1.3 provides a comprehensive overview of different types of machine learning, including their definitions and practical applications.

1.2.3 Deep learning (DL)

Deep learning, a subset of machine learning (ML), is a powerful technique within the broader field of artificial intelligence (AI) that leverages multi-layered neural networks to mimic the human brain's ability to learn and reason from data. While ML involves training algorithms on large datasets to identify patterns and make predictions, DL takes this a step further by utilizing deep neural networks to extract intricate features and relationships from complex data. This capability enables DL models to achieve state-of-the-art performance in various domains, including computer vision, natural language processing, and speech recognition [28–31]. However, DL models are computationally intensive and require significant amounts of data to train effectively. Additionally, their complex architecture can make them

Table 1.3 Types of Machine Learning [24–27]

Type of Machine Learning	*Key Features*	*Applications*
Supervised Learning	Uses labeled data to train a model to make predictions or classifications.	Image classification, spam detection, fraud detection, medical diagnosis
Unsupervised Learning	Discovers hidden patterns in unlabeled data without explicit guidance.	Customer segmentation, anomaly detection, dimensionality reduction, clustering
Semi-Supervised Learning	Semi-supervised learning leverages a hybrid approach, combining a limited quantity of labeled data with a substantial amount of unlabeled data during the training process. This technique enables the model to learn valuable patterns and insights from the unlabeled data, effectively augmenting the information derived from the labeled data. By capitalizing on the strengths of both labeled and unlabeled data, semi-supervised learning can enhance model performance and reduce the reliance on extensive labeled datasets.	Text classification, image classification, medical image analysis
Reinforcement Learning	Trains an agent to learn optimal policies by iteratively interacting with an environment, receiving rewards for positive actions and penalties for negative ones.	Game playing, robotics, autonomous vehicles, recommendation systems

challenging to interpret and debug. Table 1.4 provides a comprehensive overview of the advantages, disadvantages, limitations, and applications of DL.

Table 1.5 shows some of the most common types of deep learning architectures. There are many other variations and combinations that are used for specific tasks.

1.3 PROGRAMMING LANGUAGE AND SOFTWARE LIBRARIES FOR AI

Table 1.6 provides a general overview of popular AI libraries for different programming languages. The choice of library often depends on specific project requirements, performance needs, and developer preferences.

Table 1.4 Features of Deep Learning (DL) [32–34]

Feature	*Deep Learning (DL)*
Advantages	• DL models can automatically learn relevant features from data, reducing the need for manual feature engineering. • DL excels at processing and learning from large and complex datasets. • DL models often achieve state-of-the-art performance on challenging tasks like image and speech recognition. • DL can effectively model complex, non-linear relationships in data. • DL can work with various data types, including images, text, and numerical data. • DL enables accurate predictions for future trends and outcomes. • DL models can handle datasets with missing values. • DL is well-suited for analyzing sequential data like time series and natural language.
Disadvantages	• DL models require significant computational resources for training and inference. • DL models can be prone to over fitting, especially with limited data. • DL models can be difficult to interpret, making it challenging to understand their decision-making process. • DL models require high-quality data to perform well. • DL models may raise concerns about data privacy and security. • Developing and applying DL models often require specialized knowledge and expertise.
Limitations	• DL models generally require large amounts of data for training. • DL models can be difficult to explain and understand. • The performance of DL models can be sensitive to hyperparameter tuning.
Applications	• Image and video recognition, object detection, image segmentation. • Natural language processing (NLP), text classification, sentiment analysis, and machine translation. • Speech recognition, automatic speech recognition, speaker identification. • Predictive analytics, forecasting future trends and outcomes. • Healthcare, medical image analysis, drug discovery, disease diagnosis. • Finance, fraud detection, algorithmic trading, risk assessment. • Marketing, recommendation systems, customer segmentation, ad targeting. • Gaming, game AI, procedural content generation. • Autonomous vehicles, self-driving cars, and other autonomous vehicles

1.4 MECHANICAL DESIGN AND OPTIMIZATION USING AI

In the realm of mechanical engineering, this technology presents a powerful tool for enhancing design processes and optimizing performance. Experienced mechanical engineers have accumulated a wealth of knowledge through years of practical experience and successful projects. Machine

Table 1.5 Types of Deep Learning (DL) [35–38]

Type of Deep Learning	*Key Features*	*Applications*
Convolutional Neural Networks (CNNs)	Excellent for image and video analysis, efficiently process spatial data, utilize convolutional layers to extract features	Image classification (e.g., object recognition, facial recognition), object detection, image segmentation, medical image analysis
Recurrent Neural Networks (RNNs)	RNNs are designed to handle sequential data, including time series and text. These networks utilize recurrent connections to model temporal dependencies. To address the vanishing gradient issue, variants like Long Short-Term Memory (LSTM) and Gated Recurrent Unit (GRU) have been introduced.	Natural language processing (NLP) (e.g., language modeling, machine translation), speech recognition, Time series analysis (e.g., stock price prediction).
Generative Adversarial Networks (GANs)	Comprising a generator and a discriminator, generate realistic data samples, adversarial training process improves generation quality.	Image generation (e.g., generating realistic images of faces, objects), data augmentation, style transfer
Autoencoders	Learn efficient data representations, utilize encoder and decoder networks, used for dimensionality reduction and feature extraction	Anomaly detection, data denoising, image compression
Transformer Networks	Utilize self-attention mechanism, efficiently process long-range dependencies, widely used in NLP tasks	Natural language processing (NLP) (e.g., machine translation, text summarization), text generation, question answering.

learning offers a means to capture and leverage this expertise, enabling engineers to apply proven techniques to new challenges. By analyzing historical data, identifying patterns, and extracting insights, machine learning algorithms can help engineers make informed decisions, improve design iterations, and optimize product performance. Moreover, machine learning provides a valuable learning tool for both seasoned engineers and newcomers. By analyzing vast amounts of data, engineers can gain a deeper understanding of complex engineering phenomena, identify potential design improvements, and accelerate the learning curve. The ability to effectively

Table 1.6 Different programming language and popular libraries [39, 40]

Programming Language	*Popular Libraries*	*Highlights*
Python	TensorFlow	High-performance numerical computation, large-scale machine learning
	PyTorch	Flexible deep learning framework, dynamic computational graphs
	Scikit-learn	Machine learning in Python, simple and efficient tools for data mining and data analysis
	Keras	User-friendly neural network library, built on top of TensorFlow or Theano
	OpenCV	Open-source computer vision library, real-time image, and video processing
	NLTK	Natural Language Toolkit, platform for building Python programs to work with human language data
R	caret	Comprehensive machine learning package, tools for modeling, classification, and clustering
	tidymodels	Modern machine learning workflow integrates with the tidyverse ecosystem
	mlr	Machine learning framework offers a wide range of algorithms and methods.
Java	DeepLearning4j	Distributed deep learning framework, scalable for large datasets.
	Apache Mahout	Distributed machine learning library, scalable algorithms for large datasets.
	MLlib (Spark)	Machine learning library integrated with Apache Spark, scalable data processing and analysis.
C++	TensorFlow C++	High-performance C++ API for TensorFlow, direct hardware access.
	MXNet	Scalable deep learning framework, supports multiple programming languages
	OpenNN	Open-source neural networks library, efficient and versatile
JavaScript	TensorFlow.js	JavaScript library for machine learning, runs in the browser or Node.js
	ml5.js	Library for creating machine learning models in the browser, user-friendly API

collect, process, and interpret data is crucial for harnessing the full potential of machine learning in mechanical engineering [41–43].

1.5 PRODUCT DESIGN AND DEVELOPMENT

Machine learning is revolutionizing the design and development of products. By analyzing vast amounts of data and identifying patterns, machine learning algorithms can optimize designs, accelerate development cycles, and improve product performance.

1.6 INTEGRATED DESIGN OPTIMIZATION

Machine learning can streamline the design process by integrating various engineering disciplines, such as physics, solid mechanics, and fluid dynamics. This enables the exploration of a wider design space and the identification of optimal solutions.

- **Computer-Aided Design (CAD) Automation**: By training neural networks on large datasets of CAD models, machine learning can automate tasks like part recognition, assembly identification, and data entry, significantly speeding up the design process.
- **Generative Design**: Machine learning algorithms can generate novel and innovative designs by exploring a vast design space and identifying optimal solutions that meet specific constraints and objectives.
- **Topology Optimization**: Machine learning can be used to optimize the topology of structures, leading to lightweight and high-performance designs.
- **Robot Learning**: AI-powered robots can learn from experience and adapt to new tasks, improving their performance and versatility.
- **Predictive Maintenance**: Machine learning, powered by sensor data and historical maintenance records, can anticipate equipment failures and optimize maintenance schedules. This results in decreased downtime and increased equipment reliability.
- **Anomaly Detection**: DL models can identify unusual patterns in sensor data, indicating potential malfunctions or performance degradation.
- **Prediction of Load-Bearing Capacity**: The accurate prediction of load-bearing capacity is critical in various engineering applications, including structural engineering, civil engineering, and mechanical engineering. Traditional methods for predicting load-bearing capacity often rely on complex analytical models and experimental testing, which can be time-consuming and expensive. In recent years, artificial intelligence (AI) and machine learning (ML) techniques have emerged as powerful tools for addressing these challenges [44, 45].
- **Autonomous Systems**: ML algorithms enable autonomous systems to make decisions and respond to changes in their environment, such as self-driving vehicles or autonomous drones.
- **Accelerated Simulations**: AI can accelerate simulation processes, such as finite element analysis (FEA) and computational fluid dynamics (CFD), by optimizing simulation parameters and reducing computational time.
- **Human-Centered Design**: Recent advancements in artificial intelligence (AI), particularly machine learning, have the potential to revolutionize

the fields of design and construction. However, current AI-driven approaches often overlook the crucial role of human factors in the design process. To bridge this gap, researchers have begun to explore methods for incorporating human preferences and design strategies into AI-powered systems. By integrating AI algorithms, designer can develop more intelligent and user-centric design tools. These tools can assist designers in making informed decisions, accelerating the design process, and ultimately creating products that better meet the needs and preferences of users [46, 47].

- **Data-Driven Modeling**: ML can develop accurate predictive models for complex physical phenomena, reducing the reliance on expensive and time-consuming experiments.
- **Material Discovery and Optimization**: AI can accelerate the discovery of new materials with desired properties by analyzing large datasets of material properties and experimental results. ML can optimize material compositions and processing parameters to improve material performance.

1.7 DISCUSSION AND CONCLUSIONS

AI is revolutionizing the aerospace industry by optimizing aircraft design. By analyzing vast datasets of historical designs and simulation results, machine learning algorithms can identify patterns and trends, leading to the development of more efficient and sustainable aircraft. This enables engineers to explore a wider design space, identify optimal solutions, and reduce development time. Generative adversarial networks (GANs) are a powerful tool for accelerating product design and prototyping. By training GANs on large datasets of images or 3D models, engineers can generate realistic and diverse designs, inspiring creativity and innovation. This technology is particularly useful for creating novel and innovative product concepts. Propeller design is another area where machine learning is making a significant impact. By analyzing large datasets of historical propeller designs and performance data, machine learning algorithms can identify optimal design parameters, leading to the development of more efficient and quieter propellers. This can improve aircraft performance, reduce fuel consumption, and enhance passenger comfort. Topology optimization is a method that employs machine learning algorithms to optimize the structural design of components. By analyzing the loads and constraints imposed on a structure, these algorithms can determine the optimal material distribution, resulting in lightweight yet high-performance designs. This technology finds widespread applications in diverse industries such as aerospace, automotive, and civil engineering. By harnessing the capabilities of machine learning, engineers and designers can unlock new levels of innovation and efficiency in product development. This technology has the potential to revolutionize industries and stimulate economic growth.

REFERENCES

1. Shankar, A. (2023). Efficient data interpretation and artificial intelligence enabled IoT based smart sensing system. *Artificial Intelligence Review*, 56(12), 15053–15077.
2. Wongchai, A., Shukla, S. K., Ahmed, M. A., Sakthi, U., & Jagdish, M. (2022). Artificial intelligence-enabled soft sensor and internet of things for sustainable agriculture using ensemble deep learning architecture. *Computers and Electrical Engineering*, 102, 108128.
3. Devi, S., Garg, H., & Garg, D. (2023). A review of redundancy allocation problem for two decades: bibliometrics and future directions. *Artificial Intelligence Review*, 56(8), 7457–7548.
4. Barnawi, A., Gaba, S., Alphy, A., Jabbari, A., Budhiraja, I., Kumar, V., & Kumar, N. (2023). A systematic analysis of deep learning methods and potential attacks in internet-of-things surfaces. *Neural Computing and Applications*, 35(25), 18293–18308.
5. Verma, R. P., Pandey, K. N., & Mittal, G. (2024). Genetic-neural optimization approach for gas metal arc welding of dissimilar aluminium alloys of AA5083-O/AA6061-T6. *International Journal of Lightweight Materials and Manufacture*, 7(1), 214–220.
6. Wazid, M., Das, A. K., Chamola, V., & Park, Y. (2022). Uniting cyber security and machine learning: Advantages, challenges and future research. ICT express, 8(3), 313–321.
7. Balan, G., Arumugam, S., Muthusamy, S., Panchal, H., Kotb, H., Bajaj, M., … & Gautier, K. (2022). An improved deep learning-based technique for driver detection and driver assistance in electric vehicles with better performance. *International Transactions on Electrical Energy Systems*, 2022(1), 8548172.
8. Pande, A., Manchanda, M., Bhat, H. R., Bairy, P. S., Kumar, N., & Gahtori, P. (2022). Molecular insights into a mechanism of resveratrol action using hybrid computational docking/CoMFA and machine learning approach. *Journal of Biomolecular Structure and Dynamics*, 40(18), 8286–8300.
9. Molu, R. J. J., Tripathi, B., Mbasso, W. F., Naoussi, S. R. D., Bajaj, M., Wira, P., & Misak, S. (2024). Advancing short-term solar irradiance forecasting accuracy through a hybrid deep learning approach with Bayesian optimization. *Results in Engineering*, 23, 102461.
10. Swain, S., Bhushan, B., Dhiman, G., & Viriyasitavat, W. (2022). Appositeness of optimized and reliable machine learning for healthcare: a survey. *Archives of Computational Methods in Engineering*, 29(6), 3981–4003.
11. Singh, J., Wazid, M., Das, A. K., Chamola, V., & Guizani, M. (2022). Machine learning security attacks and defense approaches for emerging cyber physical applications: A comprehensive survey. *Computer Communications*, 192, 316–331.
12. Santhosh Kumar, H. S., & Karibasappa, K. (2022). An effective hybrid deep learning with adaptive search and rescue for brain tumor detection. *Multimedia Tools and Applications*, 81(13), 17669–17701.
13. Sharma, R., Kim, M., & Gupta, A. (2022). Motor imagery classification in brain-machine interface with machine learning algorithms: Classical approach to multi-layer perceptron model. *Biomedical Signal Processing and Control*, 71, 103101.

14. Pundir, S., Obaidat, M. S., Wazid, M., Das, A. K., Singh, D. P., & Rodrigues, J. J. (2023). MADP-IIME: malware attack detection protocol in IoT-enabled industrial multimedia environment using machine learning approach. *Multimedia Systems*, 29(3), 1785–1797.
15. Tajjour, S., Chandel, S. S., Alotaibi, M. A., Malik, H., Márquez, F. P. G., & Afthanorhan, A. (2023). Short-term solar irradiance forecasting using deep learning techniques: A comprehensive case study. *IEEE Access*, 11, 119851–119861.
16. Shahare, Y., Singh, M. P., Singh, P., Diwakar, M., Singh, V., Kadry, S., & Sevcik, L. (2023). A comprehensive analysis of machine learning-based assessment and prediction of soil enzyme activity. *Agriculture*, 13(7), 1323.
17. Nallathambi, I., Savaram, P., Sengan, S., Alharbi, M., Alshathri, S., Bajaj, M., ... & El-Shafai, W. (2023). Impact of fireworks industry safety measures and prevention management system on human error mitigation using a machine learning approach. *Sensors*, 23(9), 4365.
18. Punyavathi, R., Pandian, A., Singh, A. R., Bajaj, M., Tuka, M. B., & Blazek, V. (2024). Sustainable power management in light electric vehicles with hybrid energy storage and machine learning control. *Scientific Reports*, 14(1), 5661.
19. R. Singh, A., Kumar, R. S., Bajaj, M., Khadse, C. B., & Zaitsev, I. (2024). Machine learning-based energy management and power forecasting in grid-connected microgrids with multiple distributed energy sources. *Scientific Reports*, 14(1), 19207.
20. Tajjour, S., Chandel, S. S., Alotaibi, M. A., Malik, H., Márquez, F. P. G., & Afthanorhan, A. (2023). Short-term solar irradiance forecasting using deep learning techniques: A comprehensive case study. *IEEE Access*, 11, 119851–119861.
21. Gupta, A., Ghanshala, K., & Joshi, R. C. (2021). Machine learning classifier approach with gaussian process, ensemble boosted trees, SVM, and linear regression for 5g signal coverage mapping. *IJIMAI*, 6(6), 156–163.
22. Ojha, N., Girach, I., Sharma, K., Sharma, A., Singh, N., & Gunthe, S. S. (2021). Exploring the potential of machine learning for simulations of urban ozone variability. *Scientific reports*, 11(1), 22513.
23. Wongchai, A., Shukla, S. K., Ahmed, M. A., Sakthi, U., & Jagdish, M. (2022). Artificial intelligence-enabled soft sensor and internet of things for sustainable agriculture using ensemble deep learning architecture. *Computers and Electrical Engineering*, 102, 108128.
24. Davuluri, S. K., Alvi, S. A. M., Aeri, M., Agarwal, A., Serajuddin, M., & Hasan, Z. (2023, April). A Security Model for Perceptive 5G-Powered BC IoT Associated Deep Learning. In *2023 International Conference on Inventive Computation Technologies (ICICT)* (pp. 118–125). IEEE.
25. Tomar, V., Kumar, N., & Srivastava, A. R. (2023). Single sample face recognition using deep learning: a survey. *Artificial Intelligence Review*, 56(Suppl 1), 1063–1111.
26. Singh, T., Aherwar, A., Ranakoti, L., Bhandari, P., Singh, V., & Lendvai, L. (2023). Performance optimization of lignocellulosic fiber-reinforced brake friction composite materials using an integrated CRITIC-CODAS-based decision-making approach. *Sustainability*, 15(11), 8880.
27. Ali, S., Khan, M. S., Shah, H., Garg, H., & Alsheddy, A. (2023). BISGA: Recalculating the Entire Boolean-Valued Information System from Aggregates Using a Genetic Algorithm. *Complexity*, 2023(1), 1539563.

28. Jayaraman, R., Alshehri, M., Kumar, M., Abugabah, A., Samant, S. S., & Mohamed, A. A. (2023). Secure biomedical document protection framework to ensure privacy through blockchain. *Big data*, 11(6), 437–451.
29. Chhikara, D., Rana, S., Singh, G., Mishra, D., & Kumar, N. (2023). Blockchain-Based Partial Group Key Agreement Protocol for Intelligent Transportation Systems. *IEEE Transactions on Vehicular Technology*, 72(12), 16701–16710.
30. Wazid, M., Bera, B., Das, A. K., Mohanty, S. P., & Jo, M. (2022). Fortifying smart transportation security through public blockchain. *IEEE Internet of Things Journal*, 9(17), 16532–16545.
31. Sharma, R., Wazid, M., & Gope, P. (2021). A blockchain based secure communication framework for community interaction. *Journal of Information Security and Applications*, 58, 102790.
32. Dhiman, G., & Nagar, A. K. (2022). Blockchain-based 6G and industrial internet of things systems for industry 4.0/5.0. *Expert Systems*, 39(10), e13162.
33. Mishra, R., Ramesh, D., Bellavista, P., & Edla, D. R. (2023). Redactable Blockchain-Assisted Secure Data Aggregation Scheme for Fog-Enabled Internet-of-Farming-Things. *IEEE Transactions on Network and Service Management*, 20(4), 4652–4667.
34. Rani, S., Babbar, H., Srivastava, G., Gadekallu, T. R., & Dhiman, G. (2022). Security framework for internet-of-things-based software-defined networks using blockchain. *IEEE Internet of Things Journal*, 10(7), 6074–6081.
35. Thapliyal, S., Wazid, M., Singh, D. P., Das, A. K., Shetty, S., & Alqahtani, A. (2023). Design of robust Blockchain-envisioned authenticated key management mechanism for smart healthcare applications. *IEEE Access*, 11, 93032–93047.
36. Bansal, P., Kumar, R., Kumar, A., & Dasig Jr., D. (2024). *Book Title "Artificial Intelligence and Communication Techniques in Industry 5.0"* CRC Press, Boca Raton, FL USA, https://doi.org/10.1201/9781003494027
37. Kumar, A., Bansal, P. L., Kumar, A., Gupta, A. K., Choudhari, A.. (2024). Digital Manufacturing: Artificial Intelligence in Industry 5.0. In *Artificial Intelligence and Communication Techniques in Industry 5.0*, P. Bansal, R. Kumar, A. Kumar, D. D. Dasig, Jr. Editors, CRC Press: Boca Raton, FL, USA, Chapter 01; pp. 01–25, https://doi.org/10.1201/9781003494027-1
38. Saha, A., Ranjan, R., Yadav, A.K., Kumar, A. (2024). Payal Bansal. Multi-Objective Optimization Strategy for Enhanced Accuracy and Productivity in AWJ Machining. In *Artificial Intelligence and Communication Techniques in Industry 5.0*, P. Bansal, R. Kumar, A. Kumar, D. D. Dasig, Jr. Editors, CRC Press: Boca Raton, FL, USA, Chapter 23; pp. 360–372. https://doi.org/10.1201/9781003494027-23
39. Javeed Hussain, S., Thangavel, S., Kumar, A.. (2024). Enhancing Hospital Management through Robotics. In *Artificial Intelligence and Communication Techniques in Industry 5.0*, P. Bansal, R. Kumar, A. Kumar, D. D. Dasig, Jr. Editors, CRC Press: Boca Raton, FL, USA, Chapter 24; pp. 373–391. https://doi.org/10.1201/9781003494027-24
40. Awasthi, M. K., Kumar, A., Dutt, N., Singh, S., Editors, (2024). *Computational Fluid Flow and Heat Transfer: Advances, Design, Control & Applications*. CRC Press, Boca Raton, FL USA https://doi.org/10.1201/9781003465171
41. Awasthi, M. K., Dutt, N., Kumar, A., Hedau, A. (2024). Introduction to mathematical and computational methods. In *Computational Fluid Flow and Heat Transfer Advances, Design, Control, and Applications*, M.K. Awasthi, A.

Kumar, N. Dutt, S. Singh Editors, CRC Press: Boca Raton, FL, USA, Chapter 01; pp. 01–20. DOI: https://doi.org/10.1201/9781003465171-1

42. Awasthi, M. K., Kumar, A., & Dutt, N.. (2024). Modeling Rayleigh-Taylor instability in nanofluid layers. In *Computational Fluid Flow and Heat Transfer Advances, Design, Control, and Applications*, M.K. Awasthi, A. Kumar, N. Dutt, S. Singh Editors, CRC Press: Boca Raton, FL, USA, Chapter 13; pp. 249–262. DOI: https://doi.org/10.1201/9781003465171-13
43. Dewangan, A.K., Moinuddin, S.Q., Cheepu, M., Sajjan, S.K., Kumar, A. (2023). Thermal Energy Storage: Opportunities, Challenges and Future Scope. In *Thermal Energy Systems: Design, Computational Techniques and Applications*; Kumar, A.; Singh, V.P.; Meena, C.S. ; Dutt, N.; Editors; CRC Press: Boca Raton, FL, USA, Chapter 2; pp. 17–28. https://doi.org/10.1201/9781003395768-2
44. Tomar, P., Chauhan, P., & Kumar, A.. (2024) Anoop Kumar Shukla. *Numerical Investigation of Thermal Performance Enhancement of Solar Reservoir Using Flash Cycle, Journal of Advanced Research in Fluid Mechanics and Thermal Sciences*, 123(1), 197–221. https://doi.org/10.37934/arfmts.123.1.197221
45. Awasthi, M. K., Shukla, A.K., Kumar, A., Yadav, D., & Dutt, N. (2024). Viscous correction to the potential flow analysis of Rayleigh–Taylor instability in a Rivlin–Ericksen viscoelastic fluid layer with heat and mass transfer. *Heat Transfer* 53(6), 3072–3088. https://doi.org/10.1002/htj.23076
46. Awasthi, M. K., Dutt, N., Kumar, A., & Kumar, S.. (2024). Electrohydrodynamic capillary instability of Rivlin–Ericksen viscoelastic fluid film with mass and heat transfer. *Heat Transfer*, 53(1), 115–133. https://doi.org/10.1002/htj.22944
47. Reddy, P. N., Verma, V., Kumar, A., & Awasthi, M. (2023). CFD Simulation and Thermal Performance Optimization of Flow in a Channel with Multiple Baffles. *Journal of Heat and Mass Transfer Research*, 10(2), 257–268. https://doi.org/10.22075/JHMTR.2023.31108.1458

Chapter 2

Fundamentals of machine learning for design optimization

Ankita Pandey and Himanshu Upreti

2.1 INTRODUCTION

Conventional design optimization uses computationally costly simulations or laborious trial-and-error techniques. Conversely, machine learning (ML) offers an intriguing substitute as it can anticipate design results and investigate intricate areas far more quickly and efficiently compared to previous methods. Numerous fields, such as recommendation systems, detecting fraudulent activity, portfolio optimization, natural language processing, image and audio recognition, and automation of tasks, have applications that use this technique. For instance, recommendation systems use historical data to tailor recommendations: to suggest movies and TV shows based on reviews from users, viewing preferences, and genres; and Netflix uses content-oriented and interactive filtering. Reinforcement learning, which lets agents make decisions based on information from the context and continuously refine recommendations, significantly enhances these systems. ML impacts robots, drones, and autonomous vehicles, enhancing their adaptability to changing circumstances. This innovative approach closely aligns with data mining and data science, enabling machines to learn from data and produce accurate results.

Since the invention of computers, ML has undergone several stages of development. In 1959, "Arthur Samuel", an IBM employee and pioneer in artificial intelligence (AI) and computer games, coined the term *machine learning* (Samuel, 1959). However, in his book *The Organization of Behavior* (Milner, 1993), Canadian psychologist Donald Hebb proposed a theoretical brain structure in 1949, based on specific interactions between nerve cells. On the other hand, Warren McCulloch and Walter Pitts created the artificial neuron in 1943, a mathematical model that replicates the activity of a real neuron and is considered the first neural model ever created (McCulloch & Pitts, 1943).

According to *Time* magazine (1961), the Raytheon Company devised *Cybertron*, an experimental *learning machine* using punched tape memory that employing fundamental reinforcement learning to analyze speech patterns, sonar data, and ECGs. A human instructor or operator consistently

DOI: 10.1201/9781003589716-2

educated it to identify patterns and provided it with a correction button to compel it to analyze its erroneous conclusions. *Learning Machines* by Nilsson is a well-known work in ML, with a primary focus on ML for pattern categorization (Nilsson, 1965). A commonly used definition of the techniques studied in the field of ML was given by Tom M. Mitchell: "A computer program is said to learn from experience E with respect to some class of tasks T and performance measure P if its performance at tasks in T, as measured by P, improves with experience E" (Mitchell, 1997). This evaluation of the tasks that ML is concerned with offers a description that is fundamentally operational rather than defining the field in terms of cognition.

After that, Alan Turing proposed an article, "Computing Machinery and Intelligence", in which he replaced the question *Can machines think?* with *Can machines do what we (as thinking entities) can do*? Alan Turing, a mathematician and logician, is most recognized for his contributions to computer technology and his part in deciphering the Enigma code during WWII. His 1950 article, "Computing Machinery and Intelligence", is a seminal work in the area of AI. In this article, Turing presented the concept of machines that might replicate human intellect and behavior, laying the groundwork for future talks on machine learning (Turing, 1950).

The research environment for ML techniques includes a wide range of studies. In 1966, Joseph Weizenbaum created Eliza, a computer software capable of engaging in human conversations and extracting human-like emotions. In 1969, Arthur Bryson and Yu-Chi Ho developed a backpropagation learning method, opening the path for multi-layer ANN and laying the groundwork for deep learning, see Karjian (2024).

Due to their strong capacity to handle noisy or redundant data, extract feature variables, and create calibration models, ML approaches are now having widespread usage in many different fields, particularly in non-destructive technology and equipment intelligence. Carbonneau *et al.* (2008) considered the feasibility and comparative assessment of ML-based techniques for demand prediction by manufacturers that are biased. It has been reported that neural networks (NNs), recurrent neural networks (RNNs), and support vector machines (SVMs) performed the best when compared with traditional forecasting methods. Musumeci *et al.* (2018) highlighted that optical networking requires ML. The researcher gave an overview and an explanation of ML techniques applied in optical networking.

Abou Elassad *et al.* (2020) used different types of ML approaches to explore complex driving behavior systems. This is a review article that takes a look at 82 research papers in order to check the performance potentials of the ML model. Anaya *et al.* (2011) applied supervised and unsupervised ML methods to find out how students engage and collaborate in an online learning environment. Rashid *et al.* (2020) aimed at protecting and reducing IoT cybersecurity treats in a smart city by focusing on attack and anomaly detection techniques based on ML. In addition, they investigated ensemble techniques such as bagging, boosting, and stacking to improve the performance of the detection system.

In the last couple of years, the clinical and medical world has applied ML techniques in the detection of many diseases. Compared to the molecular diagnostic test developed traditionally, Kouchaki *et al.* (2019) found that the gradient tree boosting and logistic regression and possibly other ML approaches could significantly identify *Mycobacterium tuberculosis* (MTB). It helps in medicine resistance, decreases death rate, and controls the overflow of existing antibiotic resistance. The ML approaches were recently reviewed by EL Houby (2018), who exemplified how to use forecasting, diagnosis, and treatment of critical diseases like cancer, hepatitis, and heart-related problems using ANN, k-nearest neighbors, and decision tree methodologies.

Pandey and Kumar (2022) used the machine and deep learning technique along with its integrated architecture to distinguish between the two most common types of non-small-cell lung cancer. They developed the InceptionResNetV2 + SVM model that achieved the best performance metric out of the seven automatic integrated models. A review study based on the use of ML techniques in mental health detection was carried out by Abd Rahman *et al.* (2020). Instead of employing a typical approach, this study concentrated on offering an efficient way to analyze the retrieved data for mental health detection.

In addition to these, there are several other uses for ML techniques. A summary of current ML-based research for decision assistance in wildfire control was presented by Bot & Borges (2022). After taking into account 135 studies, they came to a conclusion that using ML techniques improves assistance throughout various stages of fire management. Bunker & Susnjak (2022) examined the accuracy attained in various sports and investigated the possibility that there is proof to back up the idea that certain sports match outcomes could be more unpredictable by nature. By developing a recommender system, Fernández-García *et al.* (2019) recommended the best elements for every user at every time.

Using intelligent data analysis techniques, they employed ML for feature selection and optimization. A summary of ML applications to ecological modelling was provided by Recknagel (2001). He discovered that ANN was helpful for ecological data visualization and nonlinear ordination. He carried out time-series modelling using Kohonen networks and recurrent networks. To identify the groundwater potential zones, Prasad *et al.* (2020) used three ML techniques in addition to an ensemble of the two best ones. This study found and classified possible zones into five types using 165 data samples and 15 contributing variables. Jung *et al.* (2021) employed a number of machine learning algorithms in injection molding manufacturing companies to maximize production quality. They looked at the elements that influence manufacturing quality using the feature selection approach. Random forests and autoencoders have the highest quality prediction results in terms of accuracy and F1-score.

Mccarthy *et al.* (2004) addressed how high-dimensional visualization and machine learning are exploratory data analysis tools that may be used to collect healthcare-relevant information from a diverse range of molecular data. Using proteome mass spectroscopy data, they identified ovarian cancer with the help of ML algorithms. Then, using gene expression data, they classified lung squamous and adenocarcinoma. A review study based on ML/DL algorithms for medical image processing was conducted by Aminizadeh *et al.* (2023). They outlined potential future tactics and different aspects of employing each strategy. They discovered that these techniques were helpful for generating decisions on vast amounts of medical data and sufficiently enabled them to support experts.

Lin *et al.* (2023) gave a brief overview of both classic and advanced ML approaches in light of the food industry's demand for quick food quality detection. According to their research, ML algorithms have shown significant promise in variety identification, categorization, and quality assessment. Using six financial domains, Nazareth & Reddy (2023) investigated how the ML methods are changing over time in financial industries. They conducted a thorough review of 126 research articles retrieved using PRISMA and identified the current state of the field and future avenues of research. Ji *et al.* (2023) presented an overview of deep learning, artificial neural networks, and other machine learning techniques. They studied the food flavor prediction and regularization of food by incorporating the molecular structure of food along with their physicochemical properties. Their study also showed that the prediction performance could be improved by the integrated design of the ML model.

Ahmed *et al.* (2024) offered an in-depth review of ML techniques applied for water resources management, targeting most recent achievements. Several fields that are subjected to the potentialities of advanced methods, such as flood control, water distribution systems, streamflow forecasts, groundwater management, and water quality, were considered in this study. The research confirmed the ability to forecast with high accuracy in various kinds of water-related sectors by using LSTM networks and hybrid machine learning approaches. Castillo-Sánchez *et al.* (2024) tracked the suicidal behavior that requires knowledge of the mental disorder that have been associated with hospital readmission. Using machine learning and conventional statistical methods, significant hospital characteristics selected adjustment disorder, alcoholism, depressive syndrome, personality disorder, and dysthymic disorder.

To find frameworks for accurate assessment, Medeiros *et al.*'s (2024) research texture descriptors on the lungs of COVID-19 patients X-rays. In the case of this study, it was found that texture descriptors improve measures for diagnosis and detection of COVID-19 by increasing predictive power and accuracy when applied with other characteristics. Modak *et al.* (2024) designed a system for the prediction of diabetes by using ML approaches. Additionally, ensemble learning was also used for boosting robustness and accuracy. The built framework was trained with python and real-world dataset. The results thus obtained indicate that the CatBoost ensemble technique was the most successful with an accuracy rate of 95.4% and a higher AUC-ROC score. More recently, a development in ML has been AutoML or automated machine learning, which improves data management and speeds up model creation. Another approach toward better utilization of edge computing in real-time processing is by having embedded machine learning, referred to as TinyML, MLOps accelerates the development, training, and deployment of ML systems. Mozumder *et al.* (2024) designed an embedded model of BERT and TF-IDF for analyzing customer satisfaction. It improved the prediction accuracy and achieved an F1 score of 0.71.

These studies provide insight into the effectiveness and versatility of ML approaches across a broad range of real-world scenarios. They underscore ML's capabilities to deal with complex problems using innovative, efficient, and data-driven solutions in a way that makes it applicable and practically useful in dealing with diverse problems. This revolutionary impact of ML is witnessed in its capabilities fir streamlines processing and decision-making in various industries including healthcare, banking, ecological modelling, and autonomous systems.

This study aspired to present a comprehensive review of machine learning (ML) by understanding its current state, its wide range of applications, its limitations, and its potential for future advancements. The overall objective of the study is to outline and explain the ML's significance in the development of contemporary technology, as well as its difficulties and the new developments that will reshape its position in addressing present-day global problems.

2.2 OVERVIEW OF MACHINE LEARNING

Creative AI, distributed companies, gaming, autonomous systems, hyperautomation, and cybersecurity are all predicted to undergo revolutionary changes due to ML. The AI industry is estimated to expand by 35% a year and reach $1.3 trillion by 2030. A branch of AI called ML is concerned with using data and algorithms to simulate human learning processes and gradually improve AI's accuracy. It enables algorithms to predict new, similar data by uncovering latent patterns in datasets, eliminating the need for explicit programming for each task. ML combines data, statistical tools, and mathematical optimization to forecast outcomes and produce actionable insights. Over the years, various experts have offered insightful definitions of ML reflecting its complexity and scope. Some of these includes:

"Machine learning is the field of study that gives computers the ability to learn without being explicitly programmed".

"Machine learning is programming computers to optimize a performance criterion using example data or past experience".

"The goal of machine learning is to develop methods that can automatically detect patterns in data, and then to use the uncovered patterns to predict future data or other outcomes of interest".

"The field of pattern recognition is concerned with the automatic discovery of regularities in data through the use of computer algorithms and with the use of these regularities to take actions".

2.2.1 Machine learning over traditional methods

In conventional programming, developers write rule-based codes based on problem statements (see Figure 2.1). It is entirely reliant on the creator's intellect. Thus, its capabilities are quite limited.

In contrast, ML focuses on using data to train an algorithm that may be used to create predictions (see Figure 2.2).

It makes predictions on fresh data using a data-driven methodology that is usually trained on past data. This makes it easier to identify trends and insights in big datasets that humans would find challenging. It comes to a conclusion that ML works better for difficult problems where a list of intricate rules or code is needed for the old way. ML can improve performance and simplify the code. Without explicit programming, ML allows

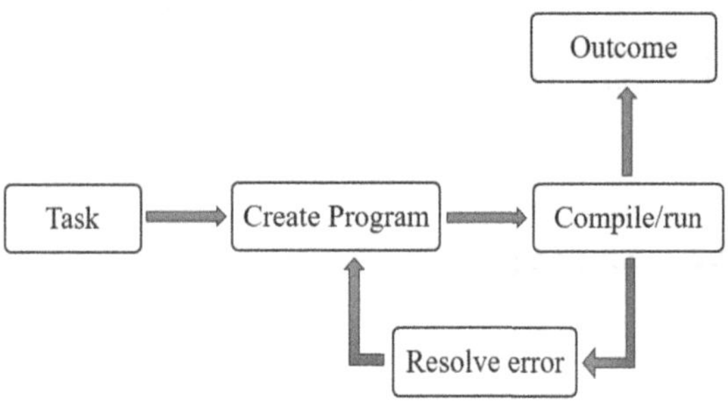

Figure 2.1 Traditional method.

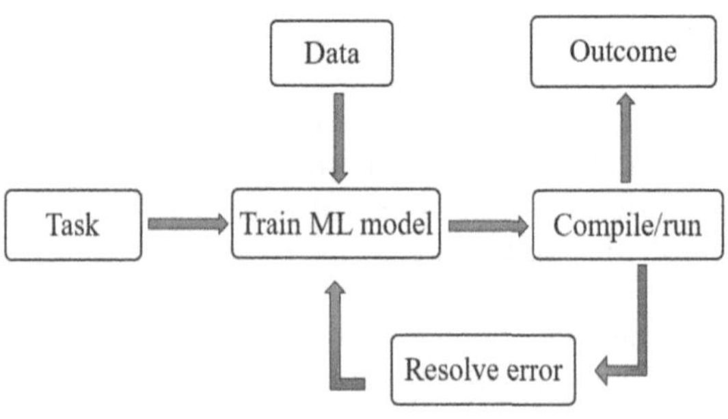

Figure 2.2 Machine learning method.

computers to acquire knowledge from data and improve their performance on specific tasks, which makes it important. ML is especially helpful for activities involving a lot of data, complicated decision-making, and dynamic contexts because of its capacity to learn from data and adjust to new circumstances.

2.2.2 Types of machine learning

In machine learning (ML), supervised, unsupervised, and reinforcement are three fundamental types of learning (see Figure 2.3), each of which solves a specific type of task. The detailed version of these learnings can be found in the following sections.

2.2.2.1 *Supervised ML*

A model is trained on the labeled dataset to categorize data or make precise predictions in supervised learning. The system uses labeled training data to learn how to map input features to targets. By learning to generalize from the input features and associated output labels, the algorithm may make predictions on new, unseen data in supervised learning. Upon receiving input data, the model modifies its weights until it is properly fitted. A matrix represents the training data in the mathematical model, whereas an array or vector is referred to as a feature vector that represents each training example. Through iterative optimization of an objective function, supervised learning creates a function that is applied to predict the output corresponding with new inputs. Categorizing spam in a different folder from your email, detecting credit card fraud, categorizing images, forecasting the weather, and identifying and classifying medical diseases are just a few of the real-world issues that supervised learning helps enterprises handle at scale.

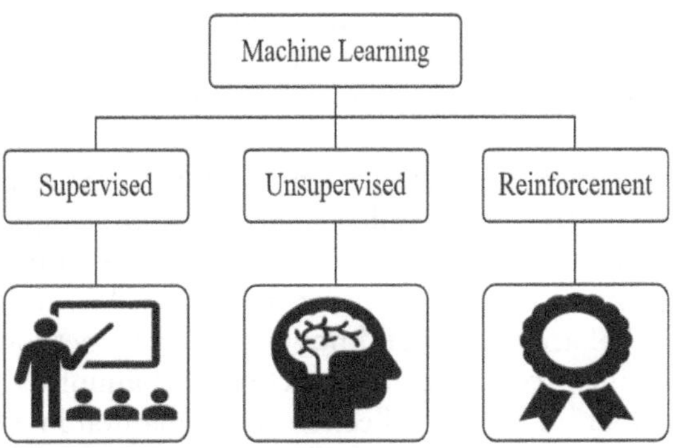

Figure 2.3 Types of machine learning.

Supervised learning can be broadly classified into two primary categories:

(i) **Regression**: This facet of supervised learning is dedicated to predicting continuous numeric outcomes using input features. It aims to uncover relationships between variables, enabling precise forecasting of values such as stock price, housing costs, or weather parameters. Advanced regression technique involved regression (linear, polynomial, and ridge), decision trees, random forests, and SVM (support vector machine), each tailored to address different complexities within datasets.

(ii) **Classification**: It entails the process of training algorithms to sort data into predefined categories-based attributes. Unlike regression, classification outputs are discrete, representing distinct classes. Binary classification algorithms provide output that belongs to one of two potential classes, while multi-class classification algorithms produce data that belongs to multiple classes. Logistic regression, Naive Bayes, decision trees, support vector machines, K-nearest neighbors and others are some of the prominent algorithms of this domain. These methodologies are found to have applications in areas such as medical diagnosis, span detection, and object recognition.

2.2.2.2 Unsupervised ML

Unsupervised learning is the process by which an algorithm learns to recognize patterns without specific training on unlabeled or labeled datasets. These algorithms find hidden patterns or data groupings within the data without human interference. The key purpose of this algorithm is to find the fundamental pattern or distribution in the input data. This approach is perfect for exploratory data analysis, cross-selling tactics, customer segmentation, anomaly detection, recommendation systems, and picture and pattern recognition because it is capable of recognizing data similarities and differences. Additionally, it is used to reduce the number of features in a model using the dimensional reduction technique.

Unsupervised learning can be divided into two primary categories:

(i) **Clustering**: Clustering methods group similar data items according to their shared attributes. Its objective is to find groupings, or clusters, of data points that are similar to one another yet different from other groups. K-means, expectation maximization, hierarchical clustering, and DBSCAN are a few well-known clustering methods.

(ii) **Dimensionality reduction**: This technique minimizes the number of input variables in a data collection while maintaining as much of the original information as feasible. This helps simplify a dataset, making it simpler to interpret and analyze. Principal Component Analysis (PCA), kernel PCA, locally linear embedding, and t-SNE are well-known dimensional reduction approaches.

2.2.2.3 *Reinforcement ML*

Reinforcement ML is similar to supervised learning, even though this method does not learn with sample data. This model uses a trial-and-error approach to learn as it goes. With reinforcement learning, an agent trains how to connect with its environment by performing activities and acquiring bonuses or retributions based on those behaviors. The objective of reinforcement learning is to learn a policy, or the conversion from states to actions, that maximizes the expected cumulative bonus over time. The optimum strategy or recommendation for a specific problem will be developed by reinforcing good outcomes. Robots, self-driving cars, AlphaGo, and AlphaStar are a few examples of reinforcement learning in action.

There are two main types of reinforcement learning:

(i) **Model-based reinforcement learning**: A model of the environment, including rewards connected to each state-action pair and the transition probabilities between states, is taught to the agent in model-based reinforcement learning. The agent then plans its actions using this model to optimize its predicted return. The Value Iteration and Policy Iteration algorithms are two well-known model-based reinforcement learning techniques.

(ii) **Model-free reinforcement learning**: Without specifically creating a model of the environment, the agent learns a policy through experience in model-free reinforcement learning. The agent engages with its surroundings and modifies its policy in response to incentives. Q-Learning, SARSA, and Deep Reinforcement Learning are a few well-known model-free reinforcement learning algorithms.

2.2.3 Advance learning

ML includes a broad range of well-known kinds of learning in addition to these core categories (Géron, 2019). Here is a quick rundown of a few of these well-liked forms.

2.2.3.1 *Semi-supervised learning*

Semi-supervised learning (SSL) offers a convenient middle ground between supervised and unsupervised learning. During training with a smaller labeled dataset, it directs the classification and extraction of attributes from a larger, unlabeled dataset. SSL can address the problem of inadequate labeled data for supervised learning (Van Engelen & Hoos, 2020). It also proves beneficial when the cost of labelling the data is extremely high. SSL features are extracted from the label and un-labeled dataset, and feature similarity is done by clustering. Based on the similarity, pseudolabels are generated for the unlabeled dataset. After that, the extracted features and corresponding labels are sent to the model for further classification (see Figure 2.4).

2.2.3.2 *Active learning*

A specific type of supervised learning called "active learning" allows a learning algorithm to interactively ask a human user or another information source questions in order to associate fresh data points with the intended results. The human user must be knowledgeable about the problem domain and capable of consulting or researching reliable sources as needed. Unlabeled data can be plentiful in some circumstances, but human labelling can be costly. In this situation, learning algorithms have the ability to actively ask the teacher or user for labels. We refer to this kind of iterative supervised learning as active learning. As the learner selects the examples, there are frequently far fewer examples needed to grasp a concept than in traditional supervised learning.

Active learning enables faster implementation of a ML system while comparing updates would necessitate a quantum or supercomputer (Novikov *et al.*, 2020). Recent advancements include multi-label active learning, hybrid active learning, and single-pass (online) active learning (Yang *et al.*, 2009; Lughofer, 2012a; Lughofer, 2012b).

2.2.3.3 *Batch learning*

In batch learning, we use the entire batch of data to train the model. It entails feeding the learning algorithm what is known as batch data, which

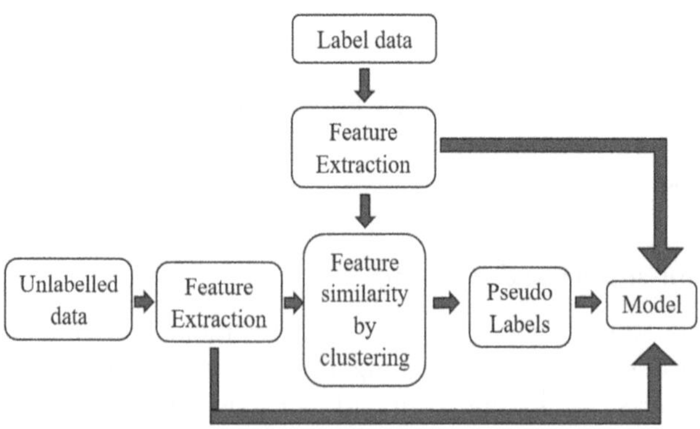

Figure 2.4 Semi-supervised learning.

enters all of the available data at once. It is less resilient to fresh data and has a poorer capacity to update its model. Once trained, the model remains unchanged until it undergoes another training phase (see Figure 2.5). It is useful in the case of stable data distributions in a context where the action is fixed (Lin & Zhou, 2017).

2.2.3.4 Online learning

It feeds data gradually, meaning that one instance or tiny batch at a time is fed in. When new data is added, it instantly cleans the dataset and is incredibly versatile. Even though it converges quickly, there are situations where we may adjust it for accuracy. Its adaptability to fresh incoming distributions demonstrates its drift-managing capabilities (see Figure 2.6). It works well for systems that update data frequently and are suited for this kind of adjustment (Duchi & Singer, 2009). Systems that continuously receive data and must adapt quickly or autonomously to changes benefit significantly from online learning. This is also a wise decision if you have limited computer power. Ren and Zhou (2024) investigated dynamic batch learning in high-dimensional sparse linear contextual bandits, in which decision-makers dynamically determine batch size and individualized action-selection methods. The fundamental learning limit is defined as a regret lower limitation and a matching upper bound, resulting in an ideal method for practical applications such as marketing and clinical trials.

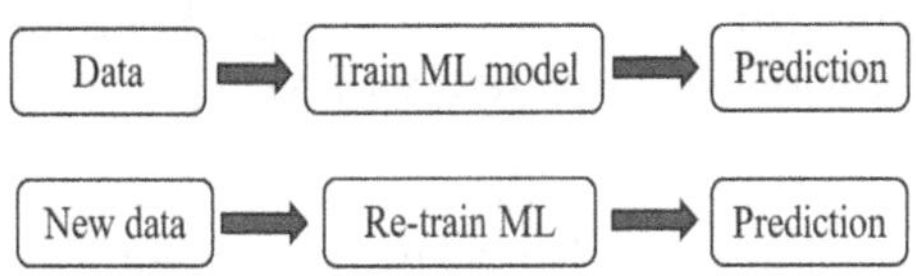

Figure 2.5 Batch learning.

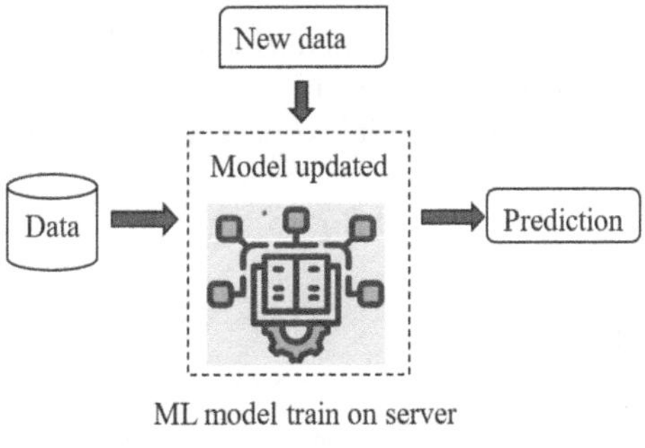

Figure 2.6 Online machine learning.

2.2.3.5 *Instance-based learning*

It is sometimes referred to as lazy learning or memory-based learning since it postpones processing until a fresh instance is classified. In this, the learning process uses the training instances to construct the hypotheses. The amount of training data determines the algorithm's temporal complexity. Every time a new query is run, the data that has already been stored is analyzed and assigned to a target function value for the new instance. It is possible to make local approximations to the goal function rather than estimating the complete dataset. The new data can be readily adapted by this learning. This kind of learning requires a significant amount of memory to store the data, and each query initiates the process of identifying a local model from the start (Aha *et al.*, 1991).

2.2.3.6 *Model-based learning*

In ML, model-based learning is a method that attempts to come up with a distinctive solution for every new problem. Its objective is to provide a unified development framework that makes it easier to create a wide range of unique models. In essence, model-based deep learning is a graphical representation of a set of assumptions. Three key concepts—factor graphs, the Bayesian worldview, and probabilistic programming—significantly came together to form this paradigm (Guo *et al.*, 2022).

2.3 CHALLENGES IN MACHINE LEARNING

Each type of ML algorithm has pros and cons. Following are some of the most prominent challenges associated with ML:

(i) The main obstacle to ML is the scarcity or broad variety of data.
(ii) A low-quality dataset with irrelevant attributes hinders a computer from learning.
(iii) For a machine to gain significant understanding, it must be heterogeneous.
(iv) Overfitting and underfitting are problems with machine learning algorithms. Overfitting occurs when the model is excessively complex compared to the training data's information and noise, while underfitting transpires when the model is too simple to grasp the data's fundamental structure.
(v) Low variation makes it hard for an algorithm to extract information.

2.4 FUTURE ASPECT

Over the past few decades, ML has transformed a number of research domains, most notably in areas like image identification, classification, and

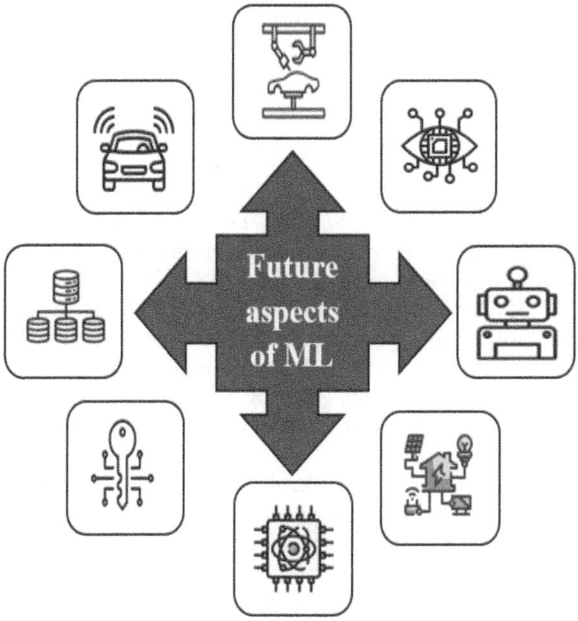

Figure 2.7 Future aspects of ML.

natural language processing (NLP). The reason for this is that ML algorithms can increase productivity and precision in tasks like data interpretation and outcome prediction. This renders them exceeding advantages for an extensive array of sectors, including manufacturing, retail, healthcare, and finance. Quantum computing holds significant promise for ML in the future and has the potential to revolutionize this sector. Faster data processing from quantum computers improves algorithms' capacity to evaluate and extract valuable insights from datasets. Machine learning's broad capabilities enable AI's unparalleled applications in business and society, affecting areas such as e-commerce, manufacturing, finance, insurance, and pharmaceuticals. Figure 2.7 shows the symbolic representation of the sectors poised for revolutionary advancements through ML.

2.5 CONCLUSION

For design optimization, ML provides strong tools that boost productivity and allow for inventive ideas in a variety of sectors. Designers may use data to make well-informed decisions and maximize performance by fusing contemporary machine learning tools with conventional design methods. Additionally, this chapter suggests advanced learning algorithms to navigate the high-dimensional and complex design space effectively. By emphasizing valuable designs and point sampling, the advanced learning procedures reduce the amount of data required.

REFERENCES

Abd Rahman, R., Omar, K., Noah, S. A. M., Danuri, M. S. N. M., & Al-Garadi, M. A. (2020). Application of machine learning methods in mental health detection: a systematic review. *IEEE Access*, 8, 183952–183964.

Abou Elassad, Z. E., Mousannif, H., Al Moatassime, H., & Karkouch, A. (2020). The application of machine learning techniques for driving behavior analysis: A conceptual framework and a systematic literature review. *Engineering Applications of Artificial Intelligence*, 87, 103312.

Aha, D. W., Kibler, D., & Albert, M. K. (1991). Instance-based learning algorithms. *Machine Learning*, 6, 37–66.

Ahmed, A. A., Sayed, S., Abdoulhalik, A., Moutari, S., & Oyedele, L. (2024). Applications of machine learning to water resources management: A review of present status and future opportunities. *Journal of Cleaner Production*, 441, 140715.

Aminizadeh, S., Heidari, A., Toumaj, S., Darbandi, M., Navimipour, N. J., Rezaei, M., ... & Unal, M. (2023). The applications of machine learning techniques in medical data processing based on distributed computing and the Internet of Things. *Computer Methods and Programs in Biomedicine*, 241, 107745.

Anaya, A. R., & Boticario, J. G. (2011). Application of machine learning techniques to analyse student interactions and improve the collaboration process. *Expert Systems with Applications*, 38(2), 1171–1181.

Bot, K., & Borges, J. G. (2022). A systematic review of applications of machine learning techniques for wildfire management decision support. *Inventions*, 7(1), 15.

Bunker, R., & Susnjak, T. (2022). The application of machine learning techniques for predicting match results in team sport: A review. *Journal of Artificial Intelligence Research*, 73, 1285–1322.

Carbonneau, R., Laframboise, K., & Vahidov, R. (2008). Application of machine learning techniques for supply chain demand forecasting. *European Journal of Operational Research*, 184(3), 1140–1154.

Castillo-Sánchez, G., Acosta, M. J., Garcia-Zapirain, B., De la Torre, I., & Franco-Martín, M. (2024). Application of machine learning techniques to help in the feature selection related to hospital readmissions of suicidal behavior. *International Journal of Mental Health and Addiction*, 22(1), 216–237

Duchi, J., & Singer, Y. (2009). Efficient online and batch learning using forward backward splitting. *The Journal of Machine Learning Research*, 10, 2899–2934.

El Houby, E. M. (2018). A survey on applying machine learning techniques for management of diseases. *Journal of Applied Biomedicine*, 16(3), 165–174.

Fernández-García, A. J., Iribarne, L., Corral, A., Criado, J., & Wang, J. Z. (2019). A recommender system for component-based applications using machine learning techniques. *Knowledge-Based Systems*, 164, 68–84.

Géron, A. (2019). *Hands-on machine learning with Scikit-Learn, Keras, and TensorFlow*. O'Reilly Media, Inc.

Guo, J., Gong, M., & Tao, D. (2022). A relational intervention approach for unsupervised dynamics generalization in model-based reinforcement learning. arXiv preprint arXiv:2206.04551.

Ji, H., Pu, D., Yan, W., Zhang, Q., Zuo, M., & Zhang, Y. (2023). Recent advances and application of machine learning in food flavor prediction and regulation. *Trends in Food Science & Technology*, 138, 738–751.

Jung, H., Jeon, J., Choi, D., & Park, J. Y. (2021). Application of machine learning techniques in injection molding quality prediction: Implications on sustainable manufacturing industry. *Sustainability*, 13(8), 4120.

Karjian, Ron. "History and evolution of machine learning: A timeline." *TechTarget*. Published June 13, 2024. Accessed November 22, 2024. . https://www.techtarget.com/whatis/feature/History-and-evolution-of-machine-learning-A-timeline

Kouchaki, S., Yang, Y., Walker, T. M., Sarah Walker, A., Wilson, D. J., Peto, T. E., … & Clifton, D. A. (2019). Application of machine learning techniques to tuberculosis drug resistance analysis. *Bioinformatics*, 35(13), 2276–2282.

Lin, J., & Zhou, D. X. (2017). Online learning algorithms can converge comparably fast as batch learning. *IEEE Transactions on Neural Networks and Learning Systems*, 29(6), 2367–2378.

Lin, Y., Ma, J., Wang, Q., & Sun, D. W. (2023). Applications of machine learning techniques for enhancing nondestructive food quality and safety detection. *Critical Reviews in Food Science and Nutrition*, 63(12), 1649–1669.

Lughofer, E. (2012a). Hybrid active learning for reducing the annotation effort of operators in classification systems. *Pattern Recognition*, 45(2), 884–896.

Lughofer, E. (2012b). Single-pass active learning with conflict and ignorance. *Evolving Systems*, 3(4), 251–271.

Mccarthy, J. F., Marx, K. A., Hoffman, P. E., Gee, A. G., O'Neil, P., Ujwal, M. L., & Hotchkiss, J. (2004). Applications of machine learning and high-dimensional visualization in cancer detection, diagnosis, and management. *Annals of the New York Academy of Sciences*, 1020(1), 239–262.

McCulloch, W. S., & Pitts, W. (1943). A logical calculus of the ideas immanent in nervous activity. *The Bulletin of Mathematical Biophysics*, 5, 115–133.

Medeiros, E. P., Machado, M. R., de Freitas, E. D. G., da Silva, D. S., & de Souza, R. W. R. (2024). Applications of machine learning algorithms to support COVID-19 diagnosis using X-rays data information. *Expert Systems with Applications*, 238, 122029.

Milner, P. M. (1993). The mind and Donald O. *Hebb. Scientific American*, 268(1), 124–129.

Mitchell, T. (1997). *Machine Learning*. McGraw Hill. p. 2. ISBN 978-0-07-042807-2.

Modak, S. K. S., & Jha, V. K. (2024). Diabetes prediction model using machine learning techniques. *Multimedia Tools and Applications*, 83(13), 38523–38549.

Mozumder, M. A. S., Nguyen, T. N., Devi, S., Arif, M., Ahmed, M. P., Ahmed, E., … & Uddin, A. (2024). Enhancing customer satisfaction analysis using advanced machine learning techniques in fintech industry. *Journal of Computer Science and Technology Studies*, 6(3), 35–41.

Musumeci, F., Rottondi, C., Nag, A., Macaluso, I., Zibar, D., Ruffini, M., & Tornatore, M. (2018). An overview on application of machine learning techniques in optical networks. *IEEE Communications Surveys & Tutorials*, 21(2), 1383–1408.

Nazareth, N., & Reddy, Y. V. R. (2023). Financial applications of machine learning: A literature review. *Expert Systems with Applications*, 219, 119640.

Nilsson N. *Learning Machines*, McGraw Hill, 1965.

Novikov, I. S., Gubaev, K., Podryabinkin, E. V., & Shapeev, A. V. (2020). The MLIP package: moment tensor potentials with MPI and active learning. *Machine Learning: Science and Technology*, 2(2), 025002.

Pandey, A., & Kumar, A. (2022, February). Deep features based automated multi-model system for classification of non-small cell lung cancer. In *2022 IEEE Delhi Section Conference (DELCON)* (pp. 1–7). IEEE.

Prasad, P., Loveson, V. J., Kotha, M., & Yadav, R. (2020). Application of machine learning techniques in groundwater potential mapping along the west coast of India. *GIScience & Remote Sensing*, 57(6), 735–752.

Rashid, M. M., Kamruzzaman, J., Hassan, M. M., Imam, T., & Gordon, S. (2020). Cyberattacks detection in iot-based smart city applications using machine learning techniques. *International Journal of Environmental Research and Public Health*, 17(24), 9347.

Recknagel, F. (2001). Applications of machine learning to ecological modelling. *Ecological Modelling*, 146(1–3), 303–310.

Ren, Z., & Zhou, Z. (2024). Dynamic batch learning in high-dimensional sparse linear contextual bandits. *Management Science*, 70(2), 1315–1342.

Samuel, A. L. (1959). Some studies in machine learning using the game of checkers. *IBM Journal of Research and Development*, 3(3), 210–229.

Time magazine (1961) Science: The Goof Button. *Time Magazine*, 78(7), August 18.

Turing, A. M. (1950). Computing Machinery and Intelligence. *Mind*, 59(236), 433–460. JSTOR

Van Engelen, J. E., & Hoos, H. H. (2020). A survey on semi-supervised learning. *Machine Learning*, 109(2), 373–440.

Yang, B., Sun, J. T., Wang, T., & Chen, Z. (2009, June). Effective multi-label active learning for text classification. In *Proceedings of the 15th ACM SIGKDD international conference on Knowledge discovery and data mining* (pp. 917–926).

Chapter 3

Case studies in machine learning for design optimization

Simisani Ndaba, Abdulhalim Musa Abubakar, Issam Ferhoune, Muhammad Tayyab Bilal, and Zakiyyu Muhammad Sarkinbaka

3.1 INTRODUCTION

Machine learning (ML), which is a branch of artificial intelligence (AI), is the process of getting systems—either computers or robots—to learn from data and assume patterns, without an explicit program. Used in design optimization, ML automates and augments decision-making processes reducing dependence on trial and error (Shrestha et al., 2021). ML techniques analyze large datasets in order to uncover insights that help improve design efficiency through performance and innovation. With modern engineering and product development complex systems requiring intelligent, adaptive solutions, ML is a critical tool. The optimization methods have undergone a thorough evolution from simple analytical techniques to highly sophisticated computational approaches. First methods were based on manual calculations and deterministic algorithms, limited with respect to the solved problems, to linear or small scale. When numerical methods and computer-based modeling became available for optimization, they evolved to solve nonlinear, multi-objective and high-dimensional problems. In today's world, ML and AI have completely changed the game of optimization with the arrival of adaptive algorithms like neural networks (NN), genetic algorithms (GA), and reinforcement learning (RL), who can run through the data and produce novel solutions (Lavor et al., 2024). AI is an essential part of modern design process, bringing in data-driven insights, automation, and intelligent decision-making. NN and support vector machines (SVM) are used with AI tools such as the support of precise modeling and simulation of design scenarios(Abbas et al., 2019; Zheng et al., 2021). AI-backed evolutionary/algorithms, optimizations, help in exploring huge design spaces, finding optimal solutions with very less human intervention. This has made AI a necessity in fields from architecture to automotive and aerospace engineering by supporting rapid prototyping and predictive maintenance and real-time performance analysis.

In several case studies that were presented, the parameters generally studied in a chemical engineering setting are reaction yields (where there are such reactions), energy efficiency, product quality, cost minimization, system

DOI: 10.1201/9781003589716-3

performance, etc. Shastri et al. (2023) and Peri (2022)analyzed the role of ML and their benefits, but they failed to pinpoint software capable of handling such tasks. Specific applications targeted are chemical reactor design, quality control in manufacturing, wind turbine placement optimization, air quality prediction, additive manufacturing enhancement, catalytic activity optimization, renewable energy systems, and environmental management (Dobbelaere et al., 2021; Gao et al., 2022; Lavor et al., 2024; Schweidtmann et al., 2021; Sharah et al., 2023). For example, GA gives chemical reactor efficiency improvements, while RL optimizes energy use in industrial processes. NNs are also used in wastewater treatment to increase water quality and shorten processing time. Therefore, ML has been widely applied in a wide range of domains from structural optimization to resource allocation, predictive maintenance to product development, and environmental conservation. The main aim of this study is to underline domain-specific case studies, especially those falling under chemical engineering. Upon examining ML applications in chemical, environmental, and product design fields, it underlines the challenges in integrating ML into workflows and provides solutions. It puts forward some important software that can be used in machine learning design optimization, among which only the familiar ones' application has been commonly reported in literature. Moreover, it offers practical solutions to problems.

3.2 FUNDAMENTALS OF MACHINE LEARNING IN DESIGN

3.2.1 Common ML algorithms for optimization

Equally important are ML-related algorithms in the improvement of design parameters through data-driven decision-making and efficiency enhancement (Peri, 2022). One of the widely used algorithms is Gradient Descent (GD), as it optimizes error reduction in optimization problems. Bayesian optimization (BO) is also popular due to its effectiveness in fine-tuning parameters. Additionally, RL is an approach greatly adopted for adaptive designs. Such algorithms find a wide range of applications in the fields of engineering structural optimization and resource allocation to production systems, where they have improved performance, reduced computational costs, and produced better results (Yang et al., 2022). NNs are one of the most powerful tools for the prediction and optimization of design outcomes. Techniques such as deep learning enable complex pattern recognition, while Convolutional Neural Network (CNNs) are applicable in design visualization and Recurrent Neural Network (RNNs) in sequential design problems. In solving difficult optimization problems in design, bio-inspired methods such as GA and Evolutionary Strategies (ES) come into prominence (Hoffmeister & Bäck, 2005). On the other hand, GA makes use of the notion of natural selection to improve solutions step by step, while ES stresses mutation and adaptation in dealing with complicated situations.

Moreso, SVMs are powerful in making decisions on design classification and prediction. They create clear boundaries between different groups, and their kernel functions help them work with data that is not straight (Berwick & Idiot, 2003). SVMs have been appreciated because they are accurate, strong against unusual data points, and able to do well even with small amounts of data (Dos Santos et al., 2021). ML-based design optimization uses deep neural networks (DNNs) because such complex, high dimensional data can be handled and good forecasting is possible in reservoir modeling and protein structure prediction. Random forests (RF) also perform strong classification and regression analyses in the presence of various inputs and has been useful in equipment health monitoring and air quality prediction by summarizing the outputs of many decision trees (DT) to achieve better accuracy rather than overfitting. Transparent, interpretable (restricted to prediction) models for predictive tasks like predicting pollution levels are achieved with DTs by breaking down data into understandable decision nodes(Hu & Li, 2017). DNNs in general, however, are good at recognizing subtle patterns, automating sophisticated processes such as species identification in biodiversity monitoring and enhancing renewable energy output—criteria necessary for deep learning (DL) to be essential in applications that need high accuracy and flexibility (Lu, 2023).

3.2.2 Computer applications for ML design optimization

Figure 3.1 showcases a wide variety of software used for ML in design optimization, categorized into groups such as general ML platforms, specialized optimization software, simulation and modeling tools, industry-specific tools, and others (Sarker, 2021). TensorFlow and PyTorch are considered the best for general-purpose ML tasks while ANSYS OptiSLang and Altair HyperStudy are powerful for sensitivity analysis and design optimization. But the most utilized across academic and industry are MATLAB, ANSYS, and Autodesk Fusion 360.

KNIME, Orange, RapidMiner, and Altair Inspire are the most user-friendly types (ideal for beginners). Despite that, StatEase Design-Expert, Tableau, and H2O.ai are the easiest to operate software. For cost-conscious users, open-source tools, such as Optuna, Apache Spark, OpenFOAM and Shogun, are recommended. The depicted ML software has been applied in numerous studies in several fields (Kaur et al., 2023), as shown in Figure 3.2.

AI and all the techniques related to it have changed the destiny of many fields, according to Shastri et al. (2023). For instance, CNNs are very good at image recognition tasks, and they're used for self-driving cars, because it helps to detect and track objects instantly in a real-time environment. CNNs are also used for medical image analysis in biomedical engineering, for

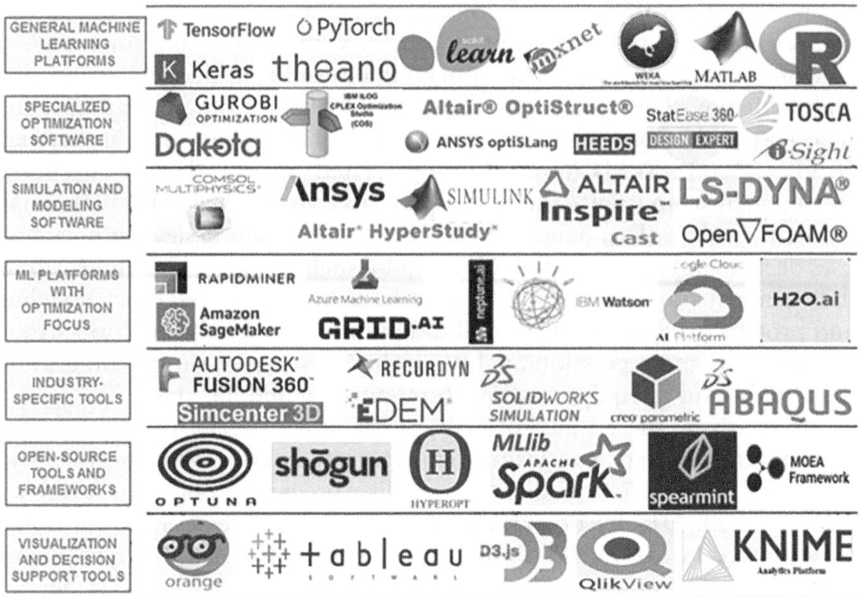

Figure 3.1 Categories of ML software for design optimization.

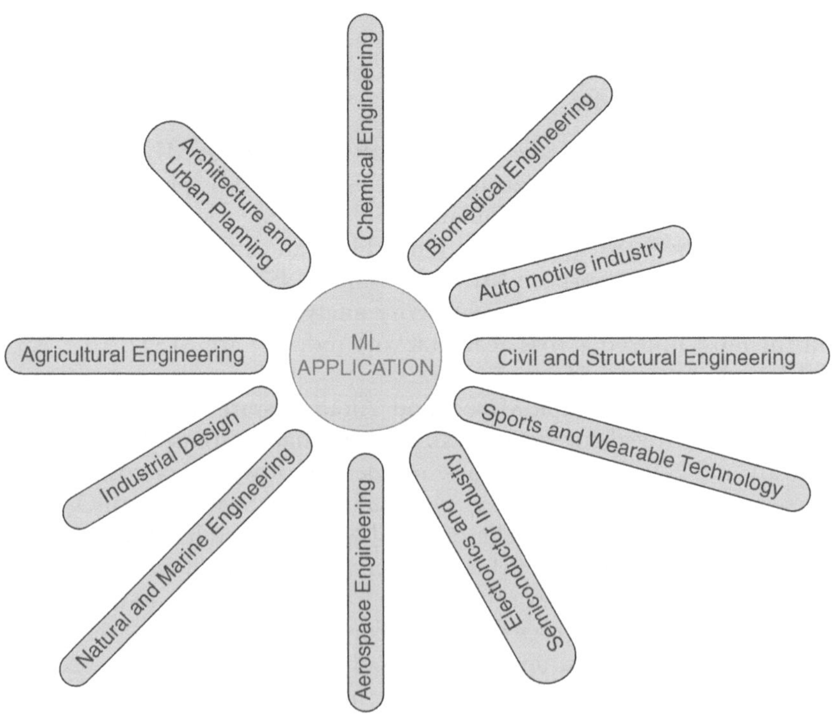

Figure 3.2 Areas of Application of ML for design optimization.

example tumor detection from MRI scans. NNs and DNNs form versatile methods for handling complex pattern recognition and predictive modeling problems, including optimizing fuel use in aerospace and marine engineering or identifying biotic stress in agriculture. BO provides a means for hyperparameters to be tuned while maintaining robust performance across a range of applications such as aerodynamic design in aerospace or structural optimization in civilengineering (Priem et al., 2020). RFs and DTs are interpretable and efficient and hence are extensively utilized in agriculture for yield prediction and in urban planning to determine environmental impact. For classification tasks, for example, structural defects in buildings, or sorting agricultural products, SVMs are used. With its role in robotics and control systems, RL is central to tasks such as autonomous navigation of vehicular systems or optimizing the energy efficiency of these systems in urban environments. The GAs inspired by natural evolution are also useful for optimization problems including design optimization in architecture and layout planning in urban development. Finally, DL extends AI's reach in broad reach by deploying the top-of-the-line capacities concerning predictive information, choice making, and real-time change among all these fields to advance its far reaching achievable and durable arrangements (Taye, 2023).

Out of several fields of study, only fourcase studies were exemplified in this study. This is due to their implementation by some multinational companies (Onu et al., 2023).

3.3 CASE STUDY 1: MACHINE LEARNING FOR CHEMICAL INDUSTRY OPERATION

Oil companies like ExxonMobil and Shell use ML models to predict the performance of oil reservoirs, optimize crude oil blending, and manage refinery operations more efficiently. Companies like Veolia use ML for optimizing water treatment processes and reducing chemical usage, while researchers are using it to develop better systems for environmental monitoring and conservation efforts. Companies like Google and Siemens are incorporating ML to predict energy demand, optimize wind turbine placement, and improve the efficiency of solar panels. In broad perspectives, its application in the chemical industry is shown in Table 3.1.

Dow Chemical and BASF are companies employing ML to predict the behavior of raw materials, optimize supply chains, and reduce waste during production.

3.4 CASE STUDY 2: MACHINE LEARNING IN PRODUCT DEVELOPMENT

General Electric (GE) and Siemens routinely use ML to optimize manufacturing processes by predicting when machines will need maintenance

Table 3.1 ML in design optimization in the chemical industry

Study	*ML technique used*	*Application area*	*Goal*	*Report*
Optimization of Chemical Reactor Design	GA	Reactor Design and Efficiency	Increased reaction yield	Rezende et al., 2008
Predicting Product Quality in Chemical Manufacturing	SVM	Quality Control and Process Monitoring	Achieve high prediction accuracy for product quality	Nachev & Stoyanov, 2012
Energy Consumption Minimization in Distillation Processes	RL	Process Optimization and Energy Efficiency	Reduced energy usage	Midgley, 2020
Predicting Oil Reservoir Behavior	DNN	Reservoir Modeling and Prediction	Improved accuracy in predicting oil recovery rates	Paroha, 2024; Temirchev et al., 2020
Catalyst Development for Polymerization Reactions	BO	Catalyst Design and Performance	Improved catalyst activity	Takasuka et al., 2024

to ensure that production lines run effortlessly. Companies like Apple, Samsung, and Sony use ML to optimize product design for features such as battery life, durability, and functionality. Table 3.2 demonstrates specific ML applications for design optimization.

Furthermore, companies like Nike, Adidas, and Under Armour utilize ML to develop products that are not only high-performing but also cost-effective. Intel, TSMC (Taiwan Semiconductor Manufacturing Company), and Samsung Electronics apply ML algorithms to improve the design of semiconductors and microchips. Basically, they predict their behavior under different operating conditions and additionally optimize power consumption.

3.5 CASE STUDY 3: ML IN ENVIRONMENTAL SCIENCE/ENGINEERING

Currently, Siemens, GE Renewable Energy, and Vestas use ML to optimize the design of wind turbines, solar panels, and energy storage systems. Smart waste management systems powered by ML (Itekari et al., 2024), namely, those implemented by Rubicon Technologies and municipal waste

Table 3.2 ML in optimizing manufacturing processes

Study	*ML approach used*	*Application area*	*Goal*	*Literature*
Predictive Maintenance for Manufacturing Equipment	RF	Equipment Health Monitoring	Reduced downtime with early fault detection	Kizito et al., 2018
Quality Control in Automated Assembly Lines	CNN	Visual Inspection and Defect Detection	Improved defect detection accuracy	Kim et al., 2023
Supply Chain Optimization in Manufacturing	RL	Supply Chain Management and Scheduling	Reduced lead times and costs	Kemmer et al., 2018
Energy Consumption Reduction in Production Lines	SVM	Process Optimization and Energy Efficiency	Lowered energy consumption	Anupong et al., 2023
Process Optimization in 3D Printing	NN and GA	Additive Manufacturing Optimization	Increased production speed and material efficiency	Shirmohammadi et al., 2021; Singh et al., 2022

departments, optimize collection routes, thereby reducing fuel consumption and operational costs. Other targeted roles of ML are presented in Table 3.3.

Researchers and organizations comparable to the Intergovernmental Panel on Climate Change (IPCC) and NASA use ML to analyze vast datasets, improve climate simulations, and predict the impact of environmental policies.

3.6 CASE STUDY 4: ML DESIGN OPTIMIZATION FOR BIOCHEMICAL ENGINEERING

Pfizer, Novartis, and Johnson & Johnson are examples of companies that use ML for drug discovery, where ML models predict molecular behaviors and interactions (Dashpute et al., 2023; Tarle et al., 2023). Consequently, they help researchers recognize promising drug candidates much sooner than traditional methods. Several other specific roles played by ML in design optimization in this field are shown in Table 3.4.

Table 3.3 Design optimization through ML in environmental management

Study	*ML method employed*	*Application area*	*Objective*	*Reference*
Predicting Air Quality Levels in Urban Areas	DT and RF	Air Quality Monitoring and Prediction	To achieve accuracy in predicting pollution levels	Hussain et al., 2021
Forest Fire Prediction and Prevention	SVM	Environmental Risk Assessment	Improved prediction accuracy	Xie & Shi, 2014
Renewable Energy Production Optimization	RL	Wind and Solar Power Generation	Increased energy efficiency	Stavrev & Ginchev, 2024
Biodiversity Monitoring in Ecosystems	DL	Ecosystem Monitoring	Automated species detection	Panigrahi et al., 2023
Wastewater Treatment Process Optimization	NN	Water Treatment and Pollution Control	Reduced processing time plus improved water quality	Hamed et al., 2004

Table 3.4 ML in bioscience/biochemical/bioengineering design optimization

Study	*ML technique activated*	*Application area*	*Goal*	*Source*
Predicting Protein Structure and Function	DL	Structural Bioinformatics	Accuracy in protein function prediction	Torrisi et al., 2020
Biomarker Discovery for Cancer Detection	SVM	Disease Diagnosis and Biomarker Discovery	Identified new biomarkers with high sensitivity	Huang et al., 2017
Optimization of Enzyme Catalysis in Biochemical Processes	GA	Enzyme Engineering	Enhanced catalytic activity	Dixit et al., 2023; Kumar et al., 2022
Drug Discovery and Design	CNN	Drug Design and Optimization	Reduced drug discovery time	Francoeur et al., 2020
Microbial Fuel Cell Optimization	RL	Bioenergy Production	Increased energy output	Shuai et al., 2024

Additionally, companies like Nestlé, PepsiCo, and Coca-Cola employ ML in the development of new flavors, enhancement of the nutritional content of food products, and the optimization of the use of ingredients(Prasad, 2024; Vyas et al., 2019).ML algorithms assist to streamline production procedures by monitoring temperature, pressure, and ingredient proportions in real time to guarantee consistency and quality.

3.7 CHALLENGES AND SOLUTIONS IN ML FOR DESIGN OPTIMIZATION

The application of ML in design optimization has delivered transformative capabilities for efficient exploration in complex design space and making data-driven decisions. However, according to Dobbelaere et al. (2021), the path to implementing ML into design workflows isn't without hurdles. The difficulties in this problem, while substantial, extend over technical, operational, and ethical dimensions, often impeding the straightforward implementation of ML for achieving the best solution. Table 3.5 describes critical challenges and their outcomes to understand the barriers.

Addressing the listed challenges in Table 3.5 requires a blend of advanced techniques, cross-disciplinary collaboration, and the development of tailored solutions for definite design applications (Schweidtmann et al., 2021). Effective collection methods to make sure that data is available and of high quality, they generate synthetic data when necessary, and cleaning data removes noise and biases, need to be used. Computational complexity fighting includes model compression and distributed computing such that calculation happens faster and more productive. Techniques such as regularization, cross-validation, and balancing model complexity can help manage the problem of overfitting; that is models that do well on training data, but not well on new data (Ghojogh & Crowley, 2019). Additionally, beyond just making ML models understandable, there is the importance of making them easier to understand, which can be done by using explainable AI (XAI) tools and visualizations to make people more confident in models' predictions (Giacobbe et al., 2023). First, Automated Machine Learning (AutoML) platforms and optimization techniques help to simplify model choices and fine tuning; second, multi-agent (MA) frameworks simplify usage of ML with traditional design methods. Adaptive algorithms and real-time learning methods keep models current and effective in changing and fast-moving environments. With encryption, secure data protocols and compliance to standards like General Data Protection Regulation(GDPR) (Goldsteen et al., 2022), as well as advanced yet secure techniques like federated learning (Brauneck et al., 2023), we can reduce security and privacy risks(El Masteri et al., 2024). Fairness aware algorithms are needed with bias and inequity; audit the datasets for diversity. For multi-objective optimization, Pareto optimization and interactive frameworks are solutions (D'Cruz et al., 1983);

Table 3.5 ML utilization challenges for design optimization

No.	*Challenge*	*Consequences*
1.	Data Availability and Quality	• Lack of sufficient, high-quality, and representative data for training ML models • Presence of noisy, incomplete, or biased datasets
2.	Computational Complexity	• High computational resources and time requirements for training complex models. • Scalability issues with large datasets or high-dimensional design spaces.
3.	Overfitting	• Risk of models performing well on training data but poorly on unseen data. • Difficulty in balancing model complexity and generalization.
4.	Interpretability and Transparency	• Challenges in understanding and explaining the predictions of black-box models like deep neural networks. • Limited user trust in decisions made by ML algorithms.
5.	Model Selection and Tuning	• Strain in choosing the most appropriate ML algorithm for specific design tasks. • Complexity of hyperparameter optimization for achieving optimal performance.
6.	Integration with Traditional Methods	• Challenges in integrating ML models with conventional design and optimization workflows. • Resistance to adoption due to reliance on traditional deterministic methods.
7.	Dynamic and Non-Stationary Environments	• Difficulty in adapting to changing design requirements or evolving data distributions. • Handling time-dependent or real-time optimization problems.
8.	Data Security and Privacy	• Risks of data breaches and ensuring the confidentiality of sensitive design information. • Compliance with data protection regulations.
9.	Ethical and Bias Concerns	• Potential biases in training data leading to unfair or suboptimal design decisions. • Ethical implications of AI-driven design choices.
10.	Multi-Objective Optimization	• Complexity in balancing trade-offs between conflicting objectives. • Trouble in defining appropriate metrics for multi-objective evaluation.
11.	Uncertainty Quantification	• Challenges in quantifying and addressing uncertainties in input data or model predictions. • Handling stochastic or probabilistic design scenarios.
12.	Limited Domain Knowledge Integration	• Difficulty in incorporating expert knowledge into ML models. • Balancing data-driven insights with domain-specific constraints.

(*Continued*)

Table 3.5 (Continued)

No.	*Challenge*	*Consequences*
13.	High Dependency on Feature Engineering	• Reliance on manual feature extraction for certain ML algorithms. • Challenges in identifying relevant features in complex design problems.
14.	Real-Time Application Constraints	• Struggle in deploying ML models for real-time decision-making or optimization. • Latency issues in time-sensitive applications.
15.	Cost and Resource Constraints	• High cost of implementing ML solutions, including hardware, software, and expertise. • Limited access to skilled personnel for developing and maintaining ML systems.
16.	Validation and Testing	• Ensuring that ML models are robust and reliable across diverse scenarios. • Lack of standardized benchmarks for evaluating model performance in design optimization.

and probabilistic modeling and ensemble methods are used for uncertainty quantification (Wang et al., 2020). Physics-informed NNs combined with expert collaboration enable integration of domain knowledge and DL alleviates manual feature engineering. The lightweight architectures and edge computing can manage real-time application constraints, whereas resource challenges can be addressed by using open-source tools, training, and collaborations (Psomakelis et al., 2023). The reliability of the model is guaranteed by standardized benchmarks and rigorous validation, allowing for greater effective and trustworthy ML-driven design optimization.

3.8 FUTURE TRENDS IN AI FOR DESIGN OPTIMIZATION

Future use of AI in design improvement encompasses increasing significance of the use of advanced technologies and innovative methods to solve complex problems. Real-time, dynamic problem solving, high-dimensional problems, and creative solutions are being unlocked by generative AI, RL, and quantum computing (Zohuri, 2023). Data-driven optimization is enabled by integration to Internet of Things (IoT) and edge computing, and digital twins and simulations help support design validation and reduce dependence on physical prototyping (Plageras & Psannis, 2022). Interpretable AI models and tools facilitating the balancing of conflicting goals are driving personal and multi-objective and sustainable designs. They are helping empower human AI collaboration and efficient and creative design workflow. Furthermore, through leveraging AI, we have reached breakthroughs in areas such as material discovery, eco-friendly design, and cross-disciplinary design (Smith-Miles, 2008). Different robustness and adaptability in complex scenarios are achieved by biologically inspired algorithms, federated learning, and advanced uncertainty quantification techniques. Augmented reality (AR) and virtual reality(VR) real-time visualization of AI-driven designs are currently enabling designs to be more interactive and user friendly and fostering trust and greater adoption (Nedbal et al., 2023). Now as the field continues to mature, benchmarks and ethical considerations are rapidly becoming critical for standardization to enhance fairness, transparency, and sustainability in AI-driven design processes.

3.9 CONCLUSION

First, this study underscores the fundamental transformative effect of ML in enhancing the design optimization across varied domains, improving the efficiency, precision, and sustainability of an undertaking. The study then explores these algorithms, such as NN, GAs, SVMs, DT, RF, DNNs, CNNs, BO, DL, and RL, by demonstrating their potential applications in manufacturing, renewable energy and environmental science, and biochemical engineering, highlighting previous studies (or fourcase study aspects). This study helps address key difficulties like information scarcity, interpretability, and multi-objective optimization, suggesting innovative answers like XAI, AutoML platforms, and crossover methodologies. But beyond this, the study not only bridges the gap of previous research, but also provides a forward-looking roadmap by integrating AR/VR and cutting-edge technology, by federated learning. A solution that addresses complex design challenges with robust prospect for scalable adoption into engineering and beyond. Future studies should look at any/all of over 30 software applications capable of solving design optimization problems in the domains highlighted herein, based on specific or multiple ML algorithms.

REFERENCES

Abbas, A. K., Al-haideri, N. A., & Bashikh, A. A. (2019). Implementing artificial neural networks and support vector machines to predict lost circulation. *Egyptian Journal of Petroleum*, *28*(4), 339–347. 10.1016/j.ejpe.2019.06.006

Anupong, W., Muda, I., AbdulAmeer, S. A., Al-Kharsan, I. H., Alviz-Meza, A., & Cárdenas-Escrocia, Y. (2023). Energy consumption and carbon dioxide production optimization in an educational building using the Supported Vector Machine and Ant Colony System. *Sustainability*, *15*(3118), 1–14. 10.3390/su15043118

Berwick, R., & Idiot, V. (2003). An idiot's guide to Support vectormachines (SVMs) (pp. 1–28). https://web.mit.edu/6.034/wwwbob/svm-notes-long-08.pdf

Brauneck, A., Schmalhorst, L., Majdabadi, M. M. K., Bakhtiari, M., Völker, U., Saak, C. C., Baumbach, J., & Baumbach, L. (2023). Federated machine learning in data-protection-compliant research. *Nature Machine Intelligence*, *5*(2–4). 10.1038/s42256-022-00601-5

D'Cruz, N., Radford, A., & Gero, J. (1983). A Pareto optimization problem formulation for building performance and design. *Engineering Optimization*, *7*(1), 17–33. 10.1080/03052158308960626

Dashpute, S. V., Pansare, J. J., Deore, Y. K., Pansare, M. J., Sonawane, P. J., Jadhav, S. P., & Patil, D. M. (2023). Artificial intelligence and machine learning in the pharmaceutical industry. *International Journal of Pharmacy and Pharmaceutical Research (IJPPR)*, *28*(2), 111–131. https://ijppr.humanjournals.com/

Dixit, M., Chhabra, D., & Shukla, P. (2023). Optimization of endoglucanase-lipase-amylase enzyme consortium from Thermomyces lanuginosus VAPS25 using Multi-Objective genetic algorithm and their bio-deinking applications. *Bioresource Technology*, *370*(128467). 10.1016/j.biortech.2022.128467

Dobbelaere, M. R., Plehiers, P. P., Van de Vijver, R., Stevens, C. V., & Van Geem, K. M. (2021). Machine Learning in Chemical Engineering: Strengths, weaknesses, Opportunities, and Threats. *Engineering*, *7*(9), 1201–1211. 10.1016/j.eng.2021.03.019

Dos Santos, R. R., De Paula Machado, T. G., & Castro, S. G. P. (2021). Support Vector Machine applied to the optimal design of composite wing panels. *Aerospace*, *8*(328). 10.3390/aerospace8110328

El Masteri, S. Z., Lenzini, G., & Demirci, H. (2024). Preserving data privacy in machine learning systems. *Computers & Security*, *137*(103605). 10.1016/j.cose.2023.103605

Francoeur, P. G., Masuda, T., Sunseri, J., Jia, A., Iovanisci, R. B., Snyder, I., & Koes, D. R. (2020). Three-dimensional Convolutional Neural Networks and a cross-docked data set for structure-based drug design. *Journal of Chemical Information and Modeling*, *60*(9). 10.1021/acs.jcim.0c00411

Gao, H., Zhu, L.-T., Luo, Z.-H., Fraga, M. A., & Hsing, I.-M. (2022). Machine learning and data science in chemical engineering. *Industrial & Engineering Chemistry Research*, *61*(24), 8357–8358. 10.1021/acs.iecr.2c01788

Ghojogh, B., & Crowley, M. (2019). The theory behind overfitting, cross validation, regularization, bagging, and boosting: Tutorial. arXiv:1905.12787, pp. 1–23. 10.48550/arXiv.1905.12787

Giacobbe, D. R., Zhang, Y., & De la Fuente, J. (2023). Explainable artificial intelligence and machine learning: novel approaches to face infectious diseases challenges. *Annals of Medicine*, *55*(2), 1–4. 10.1080/07853890.2023.2286336

Goldsteen, A., Ezov, G., & Shmelkin, R. (2022). Data minimization for GDPR compliance in machine learning models. *AI and Ethics*, *2*(9). 10.1007/s43681-021-00095-8

Hamed, M. M., Galal, M. M., & Hassanien, E. A. (2004). Prediction of wastewater treatment plant performance using Artificial Neural Networks. *Environmental Modelling & Software*, *19*(10), 919–928. 10.1016/j.envsoft.2003.10.005

Hoffmeister, F., & Bäck, T. (2005). Genetic algorithms and evolution strategies - Similarities and differences. In H. P. Schwefel&R. Männer (Eds.), *Parallel Problem Solving from Nature [PPSN 1990] Lecture Notes in Computer Science* (Vol. 496, pp. 455–469). Springer, Berlin, Heidelberg. 10.1007/BFb0029787

Hu, N., & Li, Q. (2017). Application of decision tree C4.5 algorithm in air quality evaluation. *2nd International Conference on Automation, Mechanical Control and Computational Engineering [AMCCE 2017]*, 118, 1095–1099. https://www.atlantis-press.com/article/25872429.pdf

Huang, S., Cai, N., Pacheco, P. P., Narandes, S., Wang, Y., & Xu, W. (2017). Applications of Support Vector Machine (SVM) learning in cancer genomics. *Cancer Genomics Proteomics*, *15*(1), 41–51. 10.21873/cgp.20063

Hussain, M., Afrin, S., Irin, A., & Park, S. K. (2021). Applying decision tree algorithm for air quality prediction in Bangladesh. *5th International Conference on Electrical Information and Communication Technology (EICT 2021)*. 10.1109/EICT54103.2021.9733443

Itekari, P. R., Thorushe, S. G., Kamlesh, D. K., Magdum, T. S., & Saroj, L. S. (2024). Smart waste management system. *International Research Journal of Modernization in Engineering Technology and Science (IRJMETS)*, *6*(10), 3154–3157. https://www.irjmets.com

Kaur, C., Chandel, R., Brar, T. P. S., & Sharma, S. (2023). Machine learning and its applications- A review study. *CGC International Journal of Contemporary Technology and Research*, *5*(2), 365–369. http://www.cgcijctr.com/

Kemmer, L., von Kleist, H., De Rochebouet, D., Tziortziotis, N., & Read, J. (2018). Reinforcement learning for supply chain optimization. *European Workshop on Reinforcement Learning October 2018, Lille, France*, *14*, 1–9. https://ewrl.wordpress.com/wp-content/uploads/2018/09/ewrl_14_2018_paper_44.pdf

Kim, H., Frommknecht, A., Bieberstein, B., Stahl, J., & Huber, M. F. (2023). Automated end-of-line quality assurance with visual inspection and convolutional neural networks. *Tm - Technisches Messen*, *90*(3), 196–204. 10.1515/teme-2022-0092

Kizito, R., Scruggs, P., Li, X., Kress, R., Devinney, M., & Berg, T. (2018). The application of Random Forest to predictive maintenance. *IISE Annual Conference. Proceedings; Norcross*, 354–359. https://www.researchgate.net/publication/339311213

Kumar, K., Shah, H., & Moholkar, V. S. (2022). Genetic algorithm for optimization of fermentation processes of various enzyme productions. In *Optimization of Sustainable Enzymes Production*(1st ed.,p. 24).Chapman and Hall/CRC. 10.1201/9781003292333-6

Lavor, V., De Come, F., Dos Santos, M. T., & Vianna, A. S. (2024). Machine learning in chemical engineering: Hands-on activities. *Education for Chemical Engineers*, *46*, 10–21. 10.1016/j.ece.2023.09.005

Lu, B. (2023). The application of machine learning in chemical engineering: A literature review. *ICHSSR*, *765*, 57–66. 10.2991/978-2-38476-092-3_9

Midgley, L. I. (2020). Deep Reinforcement Learning for process synthesis. *ArXiv E-Prints*, 1–11. 10.48550/arXiv.2009.13265

Nachev, A., & Stoyanov, B. (2012). Product quality analysis using Support Vector Machines. *International Journal Information Models and Analyses*, *1*, 179–192. http://www.foibg.com/ijima/vol01/ijima01-2-p10.pdf

Nedbal, C., Cerrato, C., Jahrreiss, V., Castellani, D., Pietropaolo, A., Galosi, A. B., & Somani, B. K. (2023). The role of 'artificial intelligence, machine learning, virtual reality, and radiomics' in PCNL: A review of publication trends over the last 30 years. *Therapeutic Advances in Urology*, *15*, 1–9. 10.1177/17562872231196676

Onu, P., Pradhan, A., & Mbohwa, C. (2023). The potential of industry 4.0 for renewable energy and materials development – The case of multinational energy companies. *Heliyon*, *9*(10). 10.1016/j.heliyon.2023.e20547

Panigrahi, S., Maski, P., & Thondiyath, A. (2023). Real-time biodiversity analysis using deep-learning algorithms on mobile robotic platforms. *PeerJ Computer Science*, *9*(e1502). 10.7717/peerj-cs.1502
Paroha, A. D. (2024). Predicting well performance and reservoir behaviour using Deep Neural Networks. *International Journal of Computer Trends and Technology*, *72*(3), 84–90. 10.14445/22312803/IJCTT-V72I3P112
Peri, D. (2022). Machine learning algorithms in design optimization. *ArXiv*, 1–27. 10.48550/arXiv.2203.11005
Plageras, A. P., & Psannis, K. E. (2022). Digital twins and multi-access edge computing for IIoT. *Virtual Reality & Intelligent Hardware*, *4*(6), 521–534. 10.1016/j.vrih.2022.07.005
Prasad, N. S. (2024). A study to know the impact of AI on sustainability of products in the carbonated beverages. *International Journal of Scientific Research*, *8*(3), 1–11. 10.55041/IJSREM29113
Priem, R., Gagnon, H., Chittick, I. R., Dufresne, S., Diouane, Y., & Bartoli, N. (2020). An efficient application of Bayesian optimization to an industrial MDO framework for aircraft design. *AIAA Aviation 2020 Forum*. 10.2514/6.2020-3152
Psomakelis, E., Makris, A., Tserpes, K., & Pateraki, M. (2023). A lightweight storage framework for edge computing infrastructures/EdgePersist. *Software Impacts*, *17*(100549), 1–6. 10.1016/j.simpa.2023.100549
Rezende, M. C. A. F., Costa, C. B. B., Costa, A. C., Maciel, M. R. W., & Filho, R. M. (2008). Optimization of a large scale industrial reactor by genetic algorithms. *Chemical Engineering Science*, *63*(2), 330–341. 10.1016/j.ces.2007.09.001
Sarker, I. H. (2021). Machine learning: Algorithms, real-world applications and research directions. *SN Computer Science*, *2*(160), 1–21. 10.1007/s42979-021-00592-x
Schweidtmann, A. M., Esche, E., Fischer, A., Kloft, M., Repke, J.-U., Sager, S., & Mitsos, A. (2021). Machine learning in chemical engineering: A perspective. *Chemie Ingenieur Technik*, *93*(12), 2029–2039. 10.1002/cite.202100083
Sharah, A., Owida, H., Alnaimat, F. A., & Abuowaida, S. (2023). Application of machine learning in chemical engineering: outlook and perspectives. *IAES International Journal of Artificial Intelligence (IJ-AI)*, *13*(1). 10.11591/ijai.v13.i1.pp619-630
Shastri, A. S., Shaw, K., & Singh, M. (2023). *Machine learning and optimization for engineering design*(1st ed.).SpringerSingapore. 10.1007/978-981-99-7456-6
Shirmohammadi, M., Ghoushchi, S. J., & Keshtiban, P. M. (2021). Optimization of 3D printing process parameters to minimize surface roughness with hybrid artificial neural network model and particle swarm algorithm. *Progress in Additive Manufacturing*, *6*(4), 199–215. 10.1007/s40964-021-00166-6
Shrestha, Y. R., Krishna, V., & von Krogh, G. (2021). Augmenting organizational decision-making with deep learning algorithms: Principles, promises, and challenges. *Journal of Business Research*, *123*, 588–603. 10.1016/j.jbusres.2020.09.068
Shuai, Q., Wang, Y., Jiang, Z., & Hua, Q. (2024). Reinforcement learning-based energy management for fuel cell electrical vehicles considering fuel cell degradation. *Energies*, *17*(1586). 10.3390/en17071586
Singh, B., Kumar, R., & Chohan, J. S. (2022). Multi-objective optimization of 3D Printing process using genetic algorithm for fabrication of copper reinforced

ABS parts. *Materials Today Proceedings*, *48*(Part 5), 981–988. 10.1016/j.matpr.2021.06.264

Smith-Miles, K. (2008). Cross-disciplinary perspectives on meta-learning for algorithm selection. *ACM Computing Surveys*, *41*(1), 1–23. 10.1145/1456650.1456656

Stavrev, S., & Ginchev, D. (2024). Reinforcement Learning techniques in optimizing energy systems. *Electronics*, *13*(1459), 1–17. 10.3390/electronics13081459

Takasuka, S., Ito, S., Oikawa, S., Harashima, Y., Takayama, T., Nag, A., Wakiuchi, A., Ando, T., Sugawara, T., Hatanaka, M., Miyao, T., Matsubara, T., Ohnishi, Y., Ajiro, H., & Fujii, M. (2024). Bayesian optimization of radical polymerization reactions in a flow synthesis system. *Science and Technology of Advanced Materials: Methods*. 10.1080/27660400.2024.2425178

Tarle, S., Kakad, A., & Shaikh, M. R. N. (2023). Overview: Embracing tools of artificial intelligence in pharmaceuticals. *International Journal of Science Academic Research*, *4*(6), 5749–5755.

Taye, M. M. (2023). Understanding of machine learning with deep learning: Architectures, workflow, applications and future directions. *Computers*, *12*(91). 10.3390/computers12050091

Temirchev, P., Simonov, M., Kostoev, R., Burnaev, E., Oseledets, I., Akhmetov, A., Margarit, A., Sitnikov, A., & Koroteev, D. (2020). Deep neural networks predicting oil movement in a development unit. *Journal of Petroleum Science and Engineering*, *184*(106513), 1–10. 10.1016/j.petrol.2019.106513

Torrisi, M., Pollastri, G., & Le, Q. (2020). Deep learning methods in protein structure prediction. *Computational and Structural Biotechnology Journal*, *18*, 1301–1310. 10.1016/j.csbj.2019.12.011

Vyas, S., Jain, S. S., Choudhary, I., & Chaudhary, A. (2019). Study on use of AI and Big Data for commercial system. *Amity International Conference on Artificial Intelligence [AICAI 2019]*, 1–4. 10.1109/AICAI.2019.8701361

Wang, J., Niu, T., Du, P., & Yang, W. (2020). Ensemble probabilistic prediction approach for modeling uncertainty in crude oil price. *Applied Soft Computing*, *95*(106509). 10.1016/j.asoc.2020.106509

Xie, D. W., & Shi, S. L. (2014). Prediction for burned area of forest fires based on SVM model. *Applied Mechanics and Materials*, *513–517*, 4084–4089. 10.4028/www.scientific.net/AMM.513-517.4084

Yang, J., Cheng, Z., Xiao, G., Xu, X., Wang, Y., Ding, H., & Zhou, D. (2022). Engineering design optimisation using reinforcement learning with episodic controllers. *Cognitive Computations and Systems*, *4*(4), 340–350. 10.1049/ccs2.12063

Zheng, G., Zhang, W., Zhang, W., Zhou, H., & Yang, P. (2021). Neural network and support vector machine models for the prediction of the liquefaction-induced uplift displacement of tunnels. *Underground Space*, *6*(2), 126–133. 10.1016/j.undsp.2019.12.002

Zohuri, B. (2023). Charting the future: The synergy of generative AI, quantum computing, and the transformative impact on economy, society, jobs market, and the emergence of artificial super intelligence. *Current Trends in Engineering Science*, *3*(7), 1–4. 10.54026/CTES/1050

Chapter 4

Artificial intelligence in manufacturing and production

Hemant Yadav and Shailesh Yadav

4.1 INTRODUCTION

Artificial intelligence (AI) has evolved from a futuristic idea into an essential technology shaping the manufacturing industry today. Initially, AI was mainly used for automating basic tasks, such as machine operations, but over time, it has advanced to handle much more complex functions. In its earlier stages, manufacturing relied on simple machines that could only follow pre-programmed instructions. Now, with AI, these systems can analyze large amounts of data, adapt to changing conditions, and even make decisions based on patterns and trends they observe.

AI's ability to learn from data and improve performance over time has been a game-changer for manufacturers. It allows factories to operate more efficiently, reducing waste and downtime while boosting productivity. AI systems can monitor equipment in real time, predict when a machine might need maintenance, and even suggest ways to improve production processes. This has led to smarter, more flexible production lines that can quickly respond to new demands. As a result, AI is no longer seen as an optional tool but as a fundamental part of modern manufacturing. Its integration into production processes has helped companies stay competitive in a fast-paced market, making it a critical technology for the future of the industry.

4.2 REVIEW OF LITERATURE

Since the topic Artificial intelligence is new for the world and growing day by day, this review covers research and reviews made in between 2016 and 2023 so that we can have updated ideas and views according to current scenarios. This review covers some case studies and research work as well as some reviews and by combining all, it gives a direction in which future research should go through.

Ivanov *et al.* (2016) explored the difficulties of short-term supply chain scheduling in smart factories Industry 4.0, characterized by dynamic structures, varying machine availability, and fluctuating jobs. It introduced a new

DOI: 10.1201/9781003589716-4

dynamic model and algorithm that simultaneously optimize machine structure selection and job assignments. Study combined local coordination for real-time decision-making and global optimization for the entire planning horizon, increasing efficiency. The model laid a theoretical foundation for advanced scheduling in smart factories, and future research aims to evaluate and improve self-exploration methods relative to the proposed dynamic algorithm, expanding its practical applicability.

Huoa *et al.* (2020) highlighted the critical role of both artificial intelligence (AI) and non-artificial intelligence (non-AI) in driving new product success in China's manufacturing sector. AI elements, such as technical infrastructure quality and management capabilities, coupled with non-AI aspects like emotional intelligence and responsiveness, foster innovation and operational efficiency. This research is based on data from R&D departments of manufacturing organizations, which suggested that balanced combination of AI and non-AI enhances adaptability. However, the study is limited to manufacturing firms in China, omitting cross-country and cross-industry analyses. In future, research will explore broader contexts and include more factors to improve these findings.

Ajitha and Nagra (2021) highlighted the evolving role of artificial intelligence (AI) in industries, including the automotive sector. Its structural application enables companies to enhance performance and achieve good outcomes.

The case study of Tesla Inc. exemplified AI's potential in manufacturing. Tesla's self-driving cars have received exceptional customer feedback for providing a very convenient driving experience. Elon Musk's strategic focus on direct marketing and high-speed autonomous vehicle production has successfully combined advanced technology with affordability, cementing Tesla's position as a leader in the self-driving car industry.

Arinez *et al.* (2020) explored AI applications in manufacturing, focusing on process monitoring, material property prediction, and human-robot collaboration (HRC). AI-based tools support system-level analysis, adaptive robotics, and material property modeling, offering faster and more accurate predictions compared to conventional methods. Despite varying success and challenges such as data quality, knowledge transfer, and realistic simulations, the study identifies significant opportunities for AI to transform manufacturing. Future research should address these gaps to enhance AI's role in hierarchical manufacturing processes and dynamic human-machine interactions.

Zeba *et al.* (2021) used bibliometric analysis statistical evaluations of books, papers, or other publications of academic literature to examine the expanding importance of artificial intelligence (AI) in manufacturing within the context of Industry 4.0. Important results show an increase in research after 2010. Research areas include computer engineering and science, with a growing interest in smart production, deep learning, and real-time scheduling algorithms. The study also disclosed a change in research emphasis and offers insights into new trends, laying the groundwork for further investigation into AI-driven industrial breakthroughs despite constraints like its dependence on a single database (WoSCC) and its limited search criteria.

Kim *et al.* (2022) highlighted the transformative potential of Artificial Intelligence (AI) in revolutionizing manufacturing industries, particularly in the context of Industry 4.0. While still in its infancy, AI has demonstrated superior capabilities, such as those seen in DeepMind's AlphaGo, and has become a vital tool for modeling, analysis, and automation in high-dimensional, nonlinear systems. The convergence of AI with precision engineering and manufacturing shows promise despite challenges like data scarcity, interpretability issues, and performance limitations. Emerging methodologies, including physics-informed deep learning, explainable AI, and multi-task learning, are actively addressing these hurdles. The study underscored AI's versatility, with applications spanning medical imaging, bioinformatics, and fraud detection, and envisions its impactful role in enabling future-oriented, efficient, and intelligent manufacturing solutions.

Despot *et al.* (2023) explored the changing role of artificial intelligence (AI) in the growing automotive industry, covering all the challenges like shorter development cycles and supply chain pressures. AI is revolutionizing all stages of the automotive lifecycle, from design and production to advanced features such as driver assistance and collision avoidance systems. With the help of big data and simulation-driven workflows, AI improves vehicle safety, performance, and innovation. The integration of high-performance computing and CAE tools has significantly reduced development efforts, enabling real-time optimization and pushing design boundaries. Globally, investments in AI applications highlight its potential to reshape automotive manufacturing, improve customer experiences, and foster a more efficient and innovative industry.

AI, in collaboration with complementary technologies, has emerged as a cornerstone for achieving efficiency, innovation, and sustainability in manufacturing systems.

AI's ability to emulate human intelligence enables industries to enhance decision-making processes, streamline operations, and automate repetitive tasks. In manufacturing, AI applications span design optimization, production planning, quality control, and supply chain management.

The development of smart factories in the context of Industry 4.0 demonstrates the significant influence of cyber-physical systems powered by artificial intelligence. Dynamic scheduling techniques rely on these systems to coordinate machine and job allocations in a smooth manner. When combined with optimization approaches, timing decomposition of scheduling issues enables manufacturers to adjust to changes in production settings in real time, increasing operational flexibility and efficiency.

In conclusion, AI is a force which is reshaping the manufacturing and production field with improving efficiency and sustainability by giving some additional help to the supply chain management and predictive maintenance using machine learning. In automotive vehicle designing sector, AI is helping by using different algorithms to provide predictive updates regrading failure in industries.

4.3 IMPORTANCE OF AI IN MANUFACTURING

Artificial intelligence (AI) is quickly gaining momentum in manufacturing because of the many benefits it brings. A big reason for this growth is AI's ability to reduce costs. By automating repetitive tasks and improving production workflows, AI cuts down on the need for manual labor, reduces errors, and keeps operating costs lower. This allows manufacturers to boost their production without a substantial rise in costs, giving them a major advantage in the highly competitive market today.

However, the benefits extend beyond cost savings. AI also plays a key role in enhancing efficiency at every stage of production. With real-time data analysis, AI systems enable factories to operate with greater accuracy and speed. This helps manufacturers maintain high-productivity levels while ensuring top-quality standards, ultimately making their operations more streamlined and effective. Figure 4.1 shows different roles of AI in manufacturing.

Machines with AI can keep track of their own performance, predict when they'll need maintenance, and even adjust processes automatically to get the best results. This cuts down on production delays and downtime, making the production line more streamlined and effective. Another big factor driving AI adoption is its power to turn manufacturing data into useful insights. Every day, production generates massive amounts of data, and AI can sift through it to reveal trends and opportunities. With these insights, companies

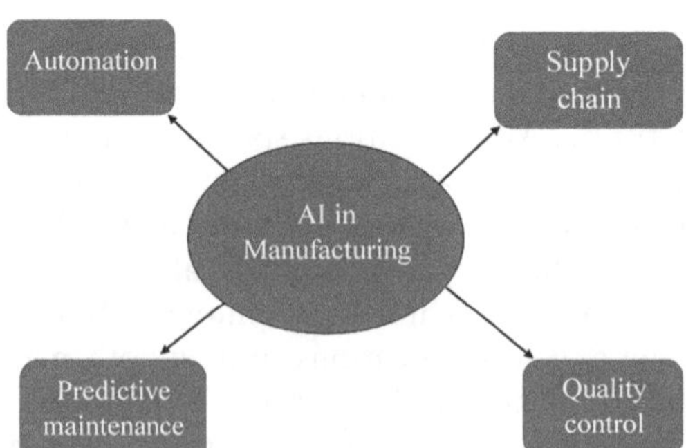

Figure 4.1 Different role of AI in manufacturing.

can predict demand, optimize supply chains, and pinpoint areas to improve. This helps them make smarter decisions, cut waste, and stay flexible to shifts in the market. Thanks to advances in AI, the idea of Industry 4.0 is becoming a reality, where AI is central to building "smart factories" (Zhong *et al.*, 2017).

AI's evolution from simple automation to intelligent, self-improving systems mark a significant shift in how products are manufactured globally (Qin *et al.*, 2016). In essence, AI is becoming indispensable in manufacturing because it streamlines operations, improves productivity, and helps businesses remain active and competitive in a dynamic industry.

4.4 PREDICTIVE MAINTENANCE USING AI

Predictive maintenance is an innovative strategy that uses artificial intelligence (AI) to keep a close eye on machinery performance in real-time, helping to spot potential failures before they happen. Machines are equipped with sensors that gather a range of data, including temperature, vibration, and pressure. This data is continuously analyzed by AI, which looks for patterns that could lead to upcoming problems, like wear and tear or the possibility of a breakdown. When these early warning signs are detected, AI can notify operators, giving them the chance to schedule maintenance at a time that works best, ultimately preventing costly downtime. By planning maintenance at the ideal time, this method helps businesses prevent unexpected equipment breakdowns, which keeps operations smooth and uninterrupted (Gaurkar *et al.*, 2021). Predictive maintenance Frame work is shown in Figure 4.2.

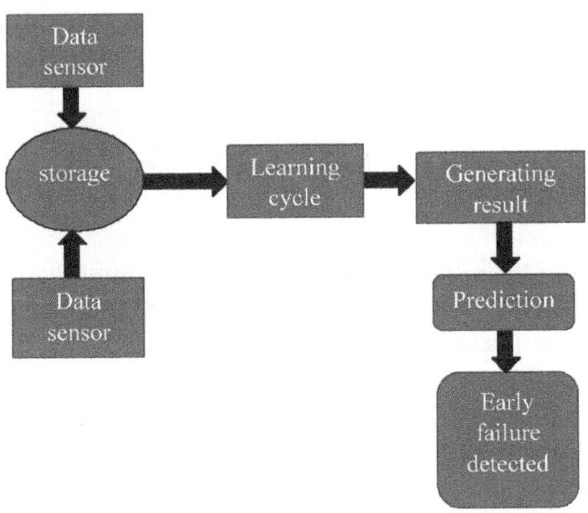

Figure 4.2 Framework of an AI-based predictive maintenance.

4.5 BENEFITS

- **Reduced downtime**: AI allows factories to schedule maintenance only when needed, avoiding unplanned disruptions and reducing the time machines are out of service, keeping production moving efficiently.
- **Reduce repair costs**: AI helps prevent costly equipment failures by identifying issues early. As a result, it saves both time and money.
- **Increased machinery lifespan**: AI-driven predictive maintenance ensures that machines receive timely upkeep and repairs before major breakdowns or failures can occur, extending their life and overall efficiency.

4.6 CASE STUDIES

Many industries are successfully using AI-driven predictive maintenance. Following are some examples of such industries.

- **Automotive Manufacturing: BMW** have adopted predictive maintenance to monitor production machinery, resulting in fewer disruptions on the assembly line and enhanced overall productivity.
- **Oil and Gas Industry**: Firms such as Shell use AI to anticipate equipment failures on oil rigs and in refineries, helping them avoid costly shutdowns and improving safety.
- **Aviation**: Boeing has implemented AI-based systems to predict maintenance needs for aircraft engines, which enhances flight reliability and brings down maintenance expenses.

These AI systems have contributed to noticeably lower unplanned downtime, reduced repair costs, and an overall improvement in production efficiency (Gaurkar *et al.*, 2021).

4.7 PROCESS OPTIMIZATION THROUGH MACHINE LEARNING

A branch of artificial intelligence (AI) called machine learning (ML) helps computers to learn from data without explicit programming. It aids systems in improving their performance over time by spotting pattern and trends and forecasting outcomes based on both previous and current data (Lee *et al.*, 2018).

In manufacturing, ML has become an invaluable tool for optimizing production processes, allowing manufacturers to make smarter decisions and to make improvements across various operations. ML algorithms analyze large datasets generated by machines, sensors, and production lines. This set of data includes information on production rates, machine performance, quality control, and more. By regularly analyzing the statistics, it helps manufacturers in spotting inefficiencies, anticipating possible problems, and not only spotting but also suggesting fixes. Through this analysis, ML enables real-time optimization of operations, ensuring that resources are used more effectively, workflows are streamlined, and energy consumption is minimized (Lee *et al.*, 2018).

4.8 OPTIMIZATION AREAS

Following are the areas which are optimized using artificial intelligence and machine learning.

- **Resource allocation**: Machine learning gives helping hands to the manufacturers in distributing resources like labor, machinery, and materials in a more efficient way. It also ensures minimum waste and optimal use.
- **Maintain Workflow**: By analyzing the flow of tasks and identifying delays or bottlenecks, ML algorithms improve production speed and reduce downtime, ensuring smoother operations.
- **Energy consumption**: ML can identify patterns in energy use and recommend strategies to reduce energy consumption, such as adjusting machine schedules or optimizing equipment performance.

4.9 IMPACT ON PRODUCTIVITY

Real-world examples highlight the impact of ML on productivity.

- **General Electric** (GE) has successfully used ML to optimize production, reducing cycle times and improving efficiency across multiple manufacturing plants.
- **Siemens** employs ML to enhance resource allocation and improve workflows, leading to reduced delays and maximized output.
- **Tesla** integrates ML into its production processes to monitor and adjust its assembly lines, reducing bottlenecks and speeding up production cycles.

The integration of AI in manufacturing processes has not only improved efficiency and resource optimization but also transformed quality control practices. As production lines become more automated and data-driven, ensuring that consistent product quality is more important than ever. AI seamlessly bridges the gap between manufacturing and quality control by incorporating intelligent systems that monitor production in real time. By optimizing workflows and detecting defects early on, AI enhances both productivity and product reliability. This integration allows manufacturers to maintain high standards of quality while scaling up production, ensuring that efficiency gains do not come at the expense of product integrity (Zhong *et al.*, 2017).

4.10 AI-ENHANCED QUALITY CONTROL

Using visual inspection, its assured quality is good for production, and in production line, every defected piece can be separated easily.

4.11 AI-BASED VISUAL INSPECTION

With a wide range of applications practically in every industry, artificial intelligence continues to be a crucial differentiator for organizations. AI is the primary enabler for next-generation technologies that are revolutionizing our way of life, from self-driving cars to Siri and Alexa. From determining which things people would enjoy best to seeing possible medical issues before they become emergencies, artificial intelligence (AI) can empower systems to make informed judgments based on historical data. Automated visual inspection is one of the many AI applications that are widely used worldwide.

Images and objects are captured, recorded, and stored on a computer. As a result, it improves efficiency and saves time. For example, if an inspector inspects an assembly line, it may take him/her a couple of hours to accomplish the inspection process, but AI-powered software will scan the assembly line within a few minutes.

Manufacturers have been able to use cloud computing and artificial intelligence with the introduction of industry 4.0 tools. In this case, a camera uploads the image on the cloud after doing a comprehensive equipment scan. Any possible flaws and nonconformities are found using a machine learning system that examines the picture for errors. Only high-quality items make it to the consumer market, because these AI systems get increasingly skilled at identifying defects by learning from each inspection cycle (Zhong *et al.*, 2017). Because AI-enhanced visual inspection detects flaws early in the production process, it increases accuracy and decreases waste.

Due to the possible mistakes that might occur during human inspection such as the loss of costly equipment, the possibility of injury, the need for rework, or the loss of a customer automated visual evaluation is necessary in many industry sectors as a high-priority job. Airport screening, food production, pharmaceuticals, and nuclear weapon manufactures are among the high-priority corporate sectors that place a premium on automated visual inspection.

4.12 AUTOMATION OF QUALITY ASSURANCE

By automated inspections that were previously done by people, AI is greatly improving the accuracy of quality assurance. Human error occurs during traditional examination, especially because of factors like subjective judgment. AI, on the other hand, continuously detects flaws and irregularities without losing accuracy.

Product quality is crucial in the manufacturing sector as flaws in products not only cause monetary losses but also diminish the company's standing. AI-powered evaluation and inspection guarantees that the manufacturing meets higher standards of norms that lead to more product sales and better client contentment. This AI-powered quality assurance tool detects faults, flaws, and other concerns that were overlooked by human inspectors. Traditionally, a quality inspector looks at visible flaws in the products. Inspection of quality by humans have limitations. Machine vision technology driven by AI allows a computer to view and comprehend pictures and employing cameras, sensors, and sophisticated algorithms to capture videos to examine the visual information. Product images are captured by cameras and sensors, and AI algorithms examine the images to find flaws like cracks, dents, and other faults. Critical product dimensions are also measured with great precision by machine vision combined with AI vision. Implementing a machine vision system includes gathering high-quality data to feed AI algorithms, pre-processing the data to include image enhancement, noise reduction, etc., it also includes training the algorithms by supplying images of both products with and without defects so that it can distinguish between them.

Automated quality assurance reduces the need for human oversight and increases inspection speed, leading to more efficient production, reduced waste, and fewer instances of product recalls or returns (Lee *et al.*, 2018). As a result, manufacturers can maintain high-quality standards while lowering costs associated with defective products.

4.13 EXAMPLES AND TECHNOLOGIES

Computer vision systems are widely used in many different fields. This is mostly because of improved image processing algorithms, more affordable manufacturing processes, and more readily available computer power. Among other places, these algorithms are utilized in virtual reality (VR), augmented reality (AR), robots, and process automation, in addition to their growing use in community service applications. As a scientific field, computer vision seeks to use algorithmic methods to extract pertinent information from pictures or image sequences for use by automated or human decision-makers. The largest problem in the field is that there are millions of pixels in a single image, and there are orders of magnitude more conceivable pixel combinations.

In electronics manufacturing, for example, computer vision can detect issues like faulty soldering or misaligned components with exceptional accuracy, preventing defective units from advancing through the production process (Qin *et al.*, 2016).

4.14 APPLICATIONS IN VARIOUS INDUSTRIES

- In **automotive industry**, like Tesla Inc., which exemplifies AI's potential in manufacturing. Tesla's self-driving cars have received exceptional customer feedback for providing a very convenient driving experience (Ajitha *et al.*, 2021).

  These autonomous vehicles utilize AI to process data from multiple sensors and cameras, enabling real-time decision-making for navigation, obstacle detection, and collision avoidance. This cutting-edge technology not only improve road safety but also provide convenience for drivers by reducing their workload and allowing them to focus on other tasks. The positive customer response to Tesla's self-driving cars underscores the transformative potential of AI in the automotive sector. By pushing the boundaries of innovation, Tesla has set a benchmark for other automobile companies that how AI can elevate both the manufacturing process and the end-user experience, paving the way for a more intelligent and connected future in the automotive industry.
- In **electronics**, AI is employed to inspect circuit boards and components, identifying microscopic defects that could affect performance. AI is redefining quality assurance in electronics manufacturing, offering a new level of precision in inspecting circuit boards and components. With the use of advanced machine learning algorithms and computer vision systems, AI can identify microscopic flaws, such as cracks, misaligned parts, soldering errors, and surface defects, which might otherwise be overlooked. Traditional inspection methods, which often depend on human expertise or basic automated systems, are prone to missing these minute imperfections or failing to maintain consistency across large production runs.
- **Pharmaceutical companies**: Pharmaceutical companies are increasingly utilizing artificial intelligence (AI) to enhance the inspection of packaging and labels, ensuring accuracy. AI-powered systems equipped with computer vision technology can analyze packaging for defects such as incorrect labeling, damaged seals, or tampered containers. Traditional inspection methods, which often involve manual checks or basic automation, are prone to human error because of human errors. AI, on the other hand, ensures a consistent and thorough inspection process, detecting even the smallest fault in text, barcodes, or expiration dates. It can also validate the information on labels against a database to confirm compliance with regulatory guidelines and product specifications.

4.15 AI FOR SUPPLY CHAIN MANAGEMENT

AI is playing an increasingly important role in supply chain management by analyzing demand patterns, inventory levels, and delivery schedules. AI-powered systems can predict future demand by analyzing past trends and external factors such as market changes or seasonality. This allows manufacturers to optimize inventory levels, avoiding both overstock and shortages. AI also streamlines delivery schedules, ensuring products to move efficiently from production to consumers while minimizing delays (Ivanov *et al.*, 2016).

4.15.1 Key benefits

- **Better demand forecasting:** AI can predict market demand with greater accuracy, allowing businesses to plan their production and inventory accordingly. This is how AI is making forecasting better on a daily basis.
- **Efficient inventory management:** AI helps manage inventory by predicting replenishment needs, reducing excess stock, and avoiding stockouts, leading to reduced cost and improved management system.
- **Reduced lead times and waste:** By optimizing supply chain processes, AI reduces production and delivery lead times while cutting down on waste due to overproduction or inefficiencies.

4.16 AI-DRIVEN LOGISTICS

AI advancements in smart logistics enable companies to optimize routes, reduce fuel consumption, and enhance delivery times. Algorithms analyze factors like traffic, weather, and vehicle performance to select the most efficient routes, improving delivery speed and reducing costs.

4.16.1 Robotics and automation powered by AI

AI is revolutionizing industrial robotics by enabling robots to perform complex tasks with precision. Traditional robots followed pre-programmed instructions, but AI-equipped robots can learn from their environment, making them capable of handling delicate operations such as precision assembly, packaging, and even quality checks. This increases accuracy and reduces the margin of error in manufacturing tasks (Lee *et al.*, 2018).

AI-powered collaborative robots, or cobots, are designed to work alongside humans safely and efficiently. These robots are equipped with sensors and AI to detect human presence and adjust their actions accordingly, making human-robot collaboration more seamless. Cobots enhance productivity by assisting workers in repetitive or physically demanding tasks while ensuring safety (Rüßmann *et al.*, 2015). Apart from their collaboration capabilities, robots provide a few other benefits. Small- and medium-sized producers now have an even playing field thanks to cobots. Large firms have benefited from automation for decades by employing conventional industrial robots. But unlike low-volume, high-mix manufacturing that is characteristic of smaller businesses, these big, costly, and sophisticated robots were made for high-volume, static production processes.

Collaborative automation has developed into a flexible, affordable, and easy-to-use technology that enables businesses of almost any size (and technical skill level) to boost output, enhance quality, and react faster to shifting client needs. It is a good technique to deal with the many company difficulties.

The rise of AI-driven automation is shifting workforce dynamics. While robots take on more repetitive and complex tasks, workers are increasingly required to manage, supervise, and program these AI systems. This transition emphasizes the need for reskilling and upskilling employees to adapt to an AI-enabled work environment (Dalenogare *et al.*, 2018).

4.16.2 AI and customization in production (flexibility in manufacturing systems)

AI is enabling manufacturers to offer mass customization, where personalized products are created at scale. AI systems analyze customer preferences and adjust production lines in real time to cater to individual needs. This level of customization allows businesses to meet diverse consumer demands without compromising on production efficiency (Rüßmann *et al.*, 2015). AI helps create flexible manufacturing systems, allowing factories to quickly adapt production lines to switch between different products. This capability enhances manufacturing agility, allowing businesses to respond to changes in demand more swiftly and efficiently (Qin *et al.*, 2016).

4.17 CHALLENGES AND ETHICAL CONSIDERATIONS

AI is new to our generation and no doubt it is giving comfort almost in all the fields, but when we are talking about taking quick and fair decisions or data sharing which should be secure, there are some fault and lack of emotions in the system, so we need to consider the ethics and human values along with the development in the era of AI, ensuring that jobs are not getting compromised too much. Following are the considerations which should be taken care of.

Data Privacy: The extensive use of AI in manufacturing raises concerns about data privacy, as large amounts of data are collected and analyzed. Ensuring that sensitive information is properly managed and protected is essential to prevent misuse or breaches (Choi *et al.*, 2018).

Workforce Displacement: AI and automation have the potential to displace certain jobs, especially those involving repetitive manual tasks. This challenge emphasizes the need for training programs that focus on developing new skills for workers to thrive in an AI-driven industry (Dalenogare *et al.*, 2018).

Bias in AI: Algorithmic bias is a growing concern in AI systems, as biased data can lead to unfair decision-making processes. In manufacturing, this could result in issues such as faulty quality assessments or imbalanced resource allocation, making it critical to ensure fairness and transparency in AI models (Choi *et al.*, 2018).

4.18 CONCLUSIONS AND FUTURE SCOPE

AI is making the manufacturing sector more efficient by enhancing quality control, supply chain management, and customization. From predictive maintenance to AI-powered robotics, the adoption of AI is enhancing productivity and driving innovation. As AI technology being advance, its impact on manufacturing and production will only grow and never decrease. The development of smarter, more sustainable systems will shape the future of the industry, enabling businesses to operate more effectively while adapting to evolving consumer needs and environmental challenges. The future of manufacturing lies in fully AI-driven systems that optimize every aspect of production, delivering greater value for both companies and consumers.

The future of manufacturing is fully automated, where factories are powered by artificial intelligence (AI) with minimal human collaborations. All production-related functions, such as resource allocation, logistics, and quality assurance, will be managed by AI in the connected systems of these factories. The future of manufacturing is envisioned as fully automated, AI-powered smart factories that require minimal human contact. These factories will use AI to handle every part of production through linked systems, from logistics and quality control to resource allocation.

AI can also play a key role in promoting sustainability by optimizing resource usage, reducing waste, and minimizing energy consumption. With growing environmental concerns, AI-driven solutions will help manufacturers reduce their environmental footprint while maintaining productivity (Zhong *et al.*, 2017).

REFERENCES

Ajitha, P. V., & Nagra, A. (2021). An overview of artificial intelligence in automobile industry–a case study on Tesla cars. *Solid State Technology*, *64*(2), 503–512.

Arinez, J. F., Chang, Q., Gao, R. X., Xu, C., & Zhang, J. (2020). Artificial intelligence in advanced manufacturing: Current status and future outlook. *Journal of Manufacturing Science and Engineering*, *142*(11), 110804.

Choi, T.-M., Wallace, S. W., & Wang, Y. (2018). Big data analytics in operations management. *Production and Operations Management*, 27(10), 1868–1883.

Dalenogare, L. S., Benitez, G. B., Ayala, N. F., & Frank, A. G. (2018). The expected contribution of industry 4.0 technologies for industrial performance. *International Journal of Production Economics*, 204, 383–394.

Despot, K., Srebrenkoska, S., & Sandeva, V. (2023). The role of artificial intelligence in automotive design. *Knowledge-International Journal*, *61*(3), 423–429.

Gaurkar, S., Kotalwar, A., & Gabale, S. (2021). Predictive maintenance of industrial machines using machine learning. *International Research Journal of Engineering and Technology*, 8(11), 515–524.

Huo, C., Hameed, J., Haq, I. U., Noman, S. M., & Chohan, S. R. (2020). The impact of artificial and non-artificial intelligence on production and operation of new products-an emerging market analysis of technological advancements a managerial perspective. *Revista Argentina de ClínicaPsicológica*, *29*(5), 69.

Ivanov, D., Dolgui, A., Sokolov, B., Werner, F., & Ivanova, M. (2016). A dynamic model and an algorithm for short-term supply chain scheduling in the smart factory industry 4.0. *International Journal of Production Research*, *54*(2), 386–402.

Kim, S. W., Kong, J. H., Lee, S. W., & Lee, S. (2022). Recent advances of artificial intelligence in manufacturing industrial sectors: A review. *International Journal of Precision Engineering and Manufacturing*, 23, 1–19.

Lee, J., Davari, H., Singh, J., & Pandhare, V. (2018). Industrial Artificial Intelligence for industry 4.0-based manufacturing systems. *Manufacturing letters*, *18*, 20–23.

Qin, J., Liu, Y., & Grosvenor, R. (2016). A categorical framework of manufacturing for industry 4.0 and beyond. *Procedia CIRP*, 52, 173–178.

Rüßmann, M., Lorenz, M., Gerbert, P., Waldner, M., Justus, J., Engel, P., & Harnisch, M. (2015). *Industry 4.0: The Future of Productivity and Growth in Manufacturing Industries*. Boston Consulting Group.

Zeba, G., Dabić, M., Čičak, M., Daim, T., & Yalcin, H. (2021). Technology mining: Artificial intelligence in manufacturing. *Technological Forecasting and Social Change*, *171*, 120971.

Zhong, R. Y., Xu, X., Klotz, E., & Newman, S. T. (2017). Intelligent Manufacturing in the Context of Industry 4.0: A Review. *Engineering*, 3(5), 616–630.

Chapter 5

Enhancing text classification through algorithmic modifications with a comparative study on AI-driven optimization techniques

Chandreyee Chakroborty, Biplab Mandal, Mohuya Byabartta Kar, and Chaahat

5.1 INTRODUCTION

It has been evidenced that in contemporary society the use of text is popular than anything else when it comes to passing information. In the modern world where technology is rapidly growing, the storage of information using computer is now more common. Therefore, most of the data which exist in the digital world are in the form of documents and files that are unstructured. As the amount of data processed textual increases due to more and more material being transmitted and stored in text documents, processing such massive volumes of textual data becomes a problem. Any type of text found on the web could include medical journals, news articles, research papers, a product review or a movie review and so on and can be classified, archived and managed using text classification algorithms. Therefore, the use of text classification provides numerous applications, such as the identification of news filtering, filtering of spam, classification of documents, detection of threats, etc.

In this study, BBC News dataset was applied, which is a popular dataset for studies on text categorization. The collection consists of 2225 documents that span five distinct categories: Entertainment, Sports, Business, Politics and Technology. Since there are many articles in each category which encompass a broad range of topics and which are similar to the respective fields, this is the most suitable dataset for the assessment of effectiveness of various text classification approaches.

Various approaches to the machine learning have been applied in the recent years to enhance text categorization efficiency. Some of the most favored used herein are Support Vector Classifier – (SVM), Convolutional Neural Networks (CNNs), Decision Tree (DT), K-Nearest Neighbors (KNN), and Naive Bayes (NB) owing to some characteristics and advantages. But for better performances while working with huge and complicated datasets, these algorithms need to be refined and optimized. Scientists have proposed a lot of modifications to such models, with the goal of improving accuracy, decreasing computational time and increasing generality. The effects of these modifications therefore have to be well understood

DOI: 10.1201/9781003589716-5

in order to decide which algorithms are most suitable for given text classification problems.

This chapter also brings out a comparison of these different text classification algorithms in terms of their effectiveness on the BBC News dataset. The chapter compares several developments and improvements that have been made in the recent works and these are discussed in relation to several models. The experiment of different algorithms is executed and the outcomes of this research guide academic and practical scholars in choosing the most desirable model for different tasks since it illustrates which approaches commonly work better than others and in what situations.

5.2 LITERATURE SURVEY

The study proposes the use of Chi-square feature selection in preprocessed data which enhances SVM text classification. The researchers also established that there is a definite advantage of stemming over lemmatization. The results of their studies are illustrated in Table 5.1 by using stemming and Chi-square feature selection; the SVM classifier captured 95% accuracy, 95% recall and 95% precision with 80% feature reduction. On the other hand, the SVM classifier depicted in Table 5.2 with lemmatization and Chi-square feature selection yielded an accuracy of 93. 24%, recall 93% and precision 93% (Haryanto *et al.*, 2018). These results suggest that with respect to lemmatization, stemming plus Chi-square feature selection significantly improves the result.

In the study, it has been found that DNN algorithms are superior to traditional methods used often in other similar tasks. Table 5.3 presents the

Table 5.1 Result after feature reduction for stemming

Feature reduction (%)	*Precision avg. (%)*	*Recall avg. (%)*	*Accuracy (%)*
0	84	81	81.08
40	91	91	90.54
80	95	95	95.05

Table 5.2 Result after feature reduction for lemmatization

Feature reduction (%)	*Precision avg. (%)*	*Recall avg. (%)*	*Accuracy (%)*
0	76	71	70.72
40	85	83	82.88
80	93	93	93.24

Table 5.3 Performance of different models

	F1 score (%)	
Categories	*LR*	*DNN*
Business	94	98
Entertainment	91	92
Politics	81	98
Sports	99	100
Technology	90	91
	Accuracy (%)	
	91	96

Table 5.4 Performance of different models

	Naïve Bayes			*SVM*		
Categories	*Accuracy (%)*	*Precision (%)*	*F Score (%)*	*Accuracy (%)*	*Precision (%)*	*F Score (%)*
Business	92.63	89.76	91.32	82.60	79.67	89.34
Entertainment	96.46	94.80	97.33	69.91	68.22	84.39
Politics	93.33	88.88	93.33	94.16	89.06	94.21
Sports	93.00	90.74	95.14	69.23	69.01	81.66
Technology	96.46	94.80	97.33	69.91	68.22	81.11

obtained results and proves that within imbalanced datasets, DNNs minimize differences in accuracy and macro-averaged F1 scores of classes. Furthermore, the efficiency is outperformed with higher mean accuracy levels and F1 scores compared to conventional methods. These observations suggest that the application of deep learning, and DNNs in particular, can enhance the efficiency and stability of classification in big data (Mauni *et al.*, 2020). As for the future work, the study suggests the practice should be continued and the enhancement of the DNN models to adopt the future big data more effectively even though this approach has drawbacks such as inherent bias and limitations.

In a study, the performance of SVM and Naïve Bayes models for sentiment analysis of news articles is evaluated. On this basis, their results are presented in Table 5.4, which indicates that Naïve Bayes has slightly better accuracy than SVM with 96.46% compared to 94.16% (Shirsat *et al.*, 2019). The process encompasses operation to clean the data which involve converting to lower case followed by stemming once the stop words, URLs, punctuations and numerals are removed. With the help of positive and

Table 5.5 Performance of different models

	Accuracy (%)		
Categories	*Random forest*	*K-NN*	*Logistic regression*
Business	93	86	99
Entertainment	90	91	98
Politics	90	99	94
Sports	97	97	99
Technology	94	89	96

Table 5.6 Performance of different models

	Accuracy (%)			
Classifers	*Lovins*	*Iterated lovins*	*Snowball*	*Null*
Naïve Bayes	96.31	95.95	96.71	96.71
Random Forest	95.77	95.82	96.31	96.00
MNBU	97.43	97.30	97.57	97.57
LogiF	96.94	97.03	96.04	96.76
SVM	96.94	97.03	96.04	97.66
Vote	97.97	97.90	97.52	97.84

negative dictionaries, they construct a Document Matrix of phrases to calculate their frequencies and define sentiment scores. As reported in previous studies, Naïve Bayes yields better results than SVM in a number of cases; it yields the highest accuracy in politics, whereas SVM yields best in sports. Of all the classifiers, Naïve Bayes is noted to perform well in entertainment and poorly in business.

The analysis of the BBC News dataset reveals that logistic regression provides tremendous accuracy in a number of categories. Surprisingly, the accuracy for the model in the business segment was estimated at 99% as captured in Table 5.5. Similarly, it provided a high level of accuracy of entertainment at 98%, political segment at 94% and sports segment of 99% (Shah *et al.* 2020). Logistic regression performed equally impressive in the technology segment also with a high accuracy of 96%. In this comparative study, among methodologies in the particular dataset, the LR model is found to be suitable for text categorization in many cases.

The researchers (Bounabi *et al.*, 2017) have presented an extensive review of various approaches to text classification. From this analysis, they point out (Table 5.6) that voting classifier is among the most stellar performers across some of the areas in the BBC News dataset. The voting classifier has been shown to consistently yield higher recognition rates compared to other

Table 5.7 Performance of different models

Models	*Accuracy (%)*	*Precision (%)*	*Recall (%)*	*F1 score (%)*
LSTM	94.6	94	94	94
CNN	98.2	98	98	98
HAN	98.2	98	98	98

methods. It is widely used for its ability to combine predictions from several independent classifiers. Analyzing the results, we can see that the Lovin+Vote method had a truly astonishing accuracy rate of 97. 97% and thus proved that it was the superior method in achieving the correct classification. Contingency matrices also known as a confusion matrix, which is described in detail by the authors, can be applied in differentiating actual and expected classes of data. This gave a good background when trying to determine the efficiency of the recognition systems in place. From this study, voting classifier is efficient and accurate in attaining high accuracy in different thematic categories of the BBC news dataset when compared with other classification techniques.

Mol and Kumar (2020) established the impacts of LSTM, CNN, and HAN models in classification of text. It was revealed that the CNN as well as the HAN models had very high levels of accuracy (Table 5.7). For the CNN model, 98.2% of accuracy was attained together with precision, recall and F1 score of 98%. The results clarify that the CNN model was impressive and competent in encoding texts through focusing on representing n-gram data patterns. Also, the HAN model attained 98.2% classification accuracy along with the precision, recall and F1 score of 98%. Despite the fact that the HAN model takes more training time than CNN, it achieved high accuracy through the application of attention mechanisms that incorporates both the sequential structure of the document and the significant components in the document. The LSTM model seemed to perform exceptionally by having a precision of 94%, recall of 94%, F1 score of 94% as well as an accuracy of 94.6%. Although it strictly adhered to the sequential structure, which allowed it to code texts effectively, its overall efficiency was lower than that of CNN and HAN models.

The study by Razali *et al.*(2020) pointed out that the multinomial Naïve Bayes classifier without word stemming is the most suitable for the classification of web pages. This study uses only textual content from websites as features, audio portions and videos are treated as non-applicable; as a result, it is possible to categorize web pages in a similar way to plain text documents. These are the characteristics on which the classifiers are trained, and the words obtained from the web pages are vectorized to create these characteristics. Before running machine learning algorithms through an instance, features usually go through some preprocessing such as stop-words removal

Table 5.8 Performance of different models

Models	*Stemming*	*Precision (%)*	*Recall (%)*	*F1 Score (%)*
Naïve Bayes	No	97	97	97
K-NN	No	96	95	96
SVM	No	98	97	97
MLP	No	97	96	96
Naïve Bayes	Yes	94	95	95
K-NN	Yes	93	92	92
SVM	Yes	95	95	95
MLP	Yes	94	93	94

Table 5.9 Performance of different models

Models	*Precision (%)*	*Recall (%)*	*F1-macro (%)*
ANN	96	96	96
Naïve Bayes	97	97	97
SVM	97	97	97

and word stemming as a way of reducing the features to be processed to an acceptable limit. But word stemming can also bring some errors in the stemmed words that may decrease the classification accuracy, especially regarding precision and recall. Notwithstanding the fact that training time has been generally reduced by stemming offers across various models, the study observes that multinomial NB without stemming performs the best for the dataset that was used in the experiments (Table 5.8).

The study by Karsana *et al.*, (2023) revealed that traditional methods for text classification include Naïve Bayes, Artificial Neural Networks (ANN) and Support Vector Machines (SVM), and each offers a specific advantage. Naïve Bayes is the most efficient for single-label classification tasks, because it is assumed that the attributes in a dataset are independent, and ANNs can be used for classification and pattern recognition but are prone to overfitting problems with datasets with multiple classifications. SVMs outperform other classifiers for binary and multiclass data, and their performances improve when the kernel is optimally chosen and dimensionality is reduced. In general, the efficiency of these methods is conditional, where SVM demonstrates high results most of the time (Table 5.9).

The Support Vector Machine (SVM) model proved to be the most accurate classifier on the BBC News dataset among the various models tested in the research (Pradhan *et al.*, 2017). This indicates that Linear SVM was significantly higher than other classifiers to be recommended as the best classifier for text classification on this dataset because of its reliability and

Table 5.10 Performance of different models

Models	*Accuracy (%)*
SVM	97.67
Naïve Bayes	96.95
K-NN	88.55
Decision Tree	79.06
Rocchio Algorithm	94.81

Table 5.11 Performance of different models

Model	*Accuracy (%)*
SVM	98.697
Random Forest	83.057
Naive Bayes	97.078
Logistic Regression	98.293

accuracy in performance (Table 5.10). Even though Naive Bayes had smaller time complexity, its accuracy was little less compared that of linear SVM. Thus, in the case of BBC News dataset, we have Linear SVM as the best classifier outperforming the accuracy and compatibility of the model with respect to this particular application.

The challenges arising from using large numbers of features in text classification is discussed, which results in time consumption and resource utilization (Smalley & Dutt, 2020). Feature selection and transformation are used to enhance classifier outcomes to solve this using dimensionality reduction methods. Some of the machine learning models that were considered included SVM, NB, RF and LR. With five-fold cross-validation, it was LR which was equally impressive with an accuracy of 98.2% (Table 5.11) obtained on BBC News dataset. NB and RF classifiers yielded lower accuracy than the SVM. Further testing showed that the SVM classifier over performed all other models and has the highest accuracy of 98.70%.

Chowdhury and Roy proposed a technique where an attention-based deep learning model first goes for training on the news corpus, and then the outcomes are compared with the existing machine learning-based models (2022). This work concentrates on supervised learning methods within the context of text categorization, particularly news categorization into sports, politics and technology. The Attention model performs better than traditional classifiers with an accuracy of 96.41% (Table 5.12), while the Multilayer Perceptron classifier proves to be 94.62% in accuracy Measures that contribute toward the Attention model include precision, recall and F1

Table 5.12 Performance of different models

Models	*Accuracy (%)*	*Precision (%)*	*Recall (%)*	*F1 score (%)*
MLP	94.62	94.72	94.62	94.61
NB	95.52	95.62	95.52	95.51
LR	95.52	95.66	95.52	95.54
RF	92.83	93.01	92.83	92.86
Attention	96.41	96.47	96.41	96.39

Table 5.13 Performance of different models

Models	*Accuracy (%)*
SVM	98.697
Logistic Regression	98.293
Naïve Bayes	97.078

score. It can be inferred that neither model is overfitting by looking at the training and validation accuracy curves, although, mis-calibration, a gradual rise in loss over time is observed here. The outcomes underline the applicability of the Attention model and indicate possible developments by the use of attention-based LSTM.

A news categorization strategy has been developed that integrates SVM and Term Frequency-Inverse Document Frequency (TF-IDF) (Dadgar *et al.*, 2016). Three primary steps are included in their approach as follows: 1) Preprocessing of the text, 2) Feature extraction (TF-IDF) and 3) Classification (SVM). It would also show an incredible precision rate of 97.84% when being tried on the BBC dataset. As the volume of unstructured data rises, text categorization as a process of identifying trends in natural language texts really gains importance. It explains how the automated method is better than manual processes and calibrates the effectiveness of the method against the older process. The SVM and TF-IDF method proves to be highly useful and accurate in the categorization of sports and news, as shown by high precision and F values.

The study proposes a semi-supervised method using Kohonen Self-organizing Maps (SOMs) with SVM, NB and DT classifiers (Barman & Chowdhury, 2020). Using a small amount of labeled data and a large number of unlabeled data in the method, Kohonen SOMs are applied to identify groups within the data. The groups are then labeled by adopting majority voting in the labeled subset. The classifiers are then trained on this labelled data. The analysis of the test results (Table 5.13) shows that the performance was much higher than random classification and that SVMs outperform NB and DT in most of the experiments. As indicated by the method, it is possible to classify a huge number of unlabeled data points with a comparatively few number of labeled data points.

Investigation has been done on some of the deep learning sequence models such as Gated Recurrent Units (GRU), Recurrent Neural Networks (RNN) and Long Short-Term Memory (LSTM) for the multiclass text classification task (Abdullahi *et al.*, 2021). Authors particularly employ the BBC news corpus for the process of theme extraction. Pretrained word embedding is used for content extraction and deep sequences are employed to retain important information. The results (Table 5.14) illustrate the performance of the sequence models where the RNN outperforms both the GRU and the LSTM and achieves an F1 score of 96.19% for the BBC dataset multiclass classification task. The evaluation suggests that deep learning sequence models are capable of reading through text inputs of reasonable lengths and dealing with natural language, although there remains potential for improvement through tuning hyperparameters.

For the purpose to evaluate sentiment in text classification, the study analyzes few machine learning models such as Naive Bayes, SVM, and Decision Trees (Chavan *et al.*, 2014). It states that SVM is superior to the other two in terms of providing accurate classification of attitudes. The study also notes that it is challenging to classify sentiment due to the subtlety of emotional indications in the text as compared to topic-based classification. In general, the work demonstrates how sentiment analysis can be automated by the machine learning algorithms as well as the processing of large volume of texts.

Text mining, machine learning, and natural language processing (NLP) play important roles in organizing and extracting meaningful insights from semi-structured and unstructured information sources. The paper (Khan *et al.*, 2010) reviews important methods and strategies used in text document classification. Information extraction, information retrieval, and the use of ontologies for knowledge representation are just a few of the topics it covers in relation to text classification. Highlighting developments in text mining and classification, pointing out enduring issues, and promoting more research into more precise and effective text classification techniques are the objectives of this study.

The multivariate Bernoulli model and the multinomial model are the first-order probabilistic algorithms which are employed for the text

Table 5.14 Performance of different models

Models	*Accuracy (%)*	*Precision (%)*	*Recall (%)*	*F1 Score (%)*
RNN	96.18	96.28	96.18	96.19
GRU	90.34	91.00	90.34	90.16
LSTM	91.46	91.49	91.46	91.44

categorization and both the types of models use the naive Bayes assumption (McCallum & Nigam, 1998). If the parameters of the multivariate Bernoulli model is set properly when the vocabulary is small, then the multinomial model would usually perform even better than the latter one. Specifically, the multinomial model reduces the error rate on the average by 27%, web pages, UseNet articles, and Reuters newswire items as compared to the multivariate Bernoulli model. The paper discusses in detail how these models work with a note that the multinomial models are in general more efficient in text classification tasks which involve large numbers of vocabularies.

The paper proposes an analysis of automated text classification and intrusion detection based on a ML approach. Labeled datasets such as Reuters and Movie Review corpora evaluate different classifiers such as NB, LR, Stochastic-GD, SVM and ANNs (Kumar *et al.*, 2019). The paper affirms that machine learning is crucial in dealing with substantial documents and fortifying computer network through Intrusion Detection Systems (IDS). IDS can incorporate the use of anomaly detection for new attack detection or misuse detection for already known threats. The paper also highlights on the difficulties of the genre-based textual classification as well as IDS requirements for hybrid detection methods.

To enhance decoding in statistical machine translation systems, the research proposes a machine learning approach. It points out the interests of decoding speed/ translation accuracy trade-off and devotes a great deal of consideration to discuss the use of threshold and histogram pruning (Banik *et al., 2019*). Instead of using specific values, the authors propose the approach of changing the beam threshold and stack size for the attributes and organization of the text. It can be seen from the experiments that higher decoding speed and translation accuracy are achieved on ASLs such as English-Hindi, Hindi-Konkani, Bengali-Hindi, and Hindi-Marathi. To accelerate decoding, the method optimizes cube pruning, an equally important part of the code.

The research is concentrated on the machine learning-based text categorization (TC) and reviewing classifiers including the Rocchio classifier and SVM (Luo, 2021). Testing it on 1033 English text documents, it reports that SVM works satisfactorily, especially, when the feature sets are more (greater than 4000) with a classification percentage of around 90%. In this paper, the performance of the Hybrid methods, Neural Networks, DT, and NB Algorithms is compared. This strengthens the argument of the significance of preprocessing operations like stemming and stop-word elimination. The importance of the machine learning technique for processing large text corpora is also discussed in the article and its applications in opinion mining, spam filtering, and fake news detection. As for the above results, the overall accuracy of the classification task in SVM is higher than that of Naïve Bayes.

The study aims at evaluating different supervised machine learning algorithms including ANN-Back Propagation Network, Support Vector

Machine-SVC and Linear SVC and Stochastic Gradient Descent (Mishu & Rafiuddin, 2016). It outlines the process of text classification, including document reading, tokenization of texts, stemming, stop words elimination, converting text to vectors, feature selection, use of supervised learning methods and accuracy determination. Several methods are tested as a way of comparing many classifiers and the results are measured using F1 score.

This study focuses mainly on the text classification, which is the process of categorizing documents into predefined classes (Bhattacharyya *et al.*, *2023*). This method is particularly important when it comes to email filtering, search engines, web content and many others. Some of the issues associated with text categorization include noisy data, irrelevant or redundant feature and or high dimensionality. To solve these problems and increase classification performance, there are feature reduction which includes feature transformation and selection. This research also compares and defines several approaches to text categorization and feature extraction such as feature transformation and filter methods of feature selection. It also describes and compares machine learning methods for text classification. Lastly, an overview of the process of text categorization along with the assessment measuring tools is presented in the article.

This research outlines a machine translation and summary evaluation metrics model for Recognizing Textual Entailment (RTE) (Agarwal & Mittal, 2012). To determine the entailment link between the two texts, the authors examine the effectiveness of the metrics commonly used in MT and other tasks like BLEU, METEOR, and ROUGE among others. It compares these metrics for suitability in textual entailment tasks experimentally on RTE datasets and reports promising outcomes. The article is focused on demonstrating the application of TE in domains such as MT, question answering, and summarization, and it is beneficial for handling the language unpredictability, which is one of the challenges in NLP.

This chapter focuses on the challenges of classifying a large number of documents. It provides the metrics and tests the efficiency of the raw data and the processed data with few traditional classification models (Kamath *et al.*, 2018). It also considers the effectiveness of deep learning techniques especially CNN with respect to other traditional methods. Two datasets were employed in the study: tobacco dataset in public domain and another dataset from an insurance company where they divided classes hierarchically. Overall, conventional techniques gave higher accuracy of results, while CNNs had better results on both datasets compared to Logistic Regression. Traditional methods gave better results with the raw images while the CNNs gave better results with processed images.

In this work, text categorization performance is compared using the machine learning techniques – SVM, k-NN, LR, Multinomial-NB, and RF. In the IMDB dataset, k-NN provided a higher accuracy than both SVM and LR, while in the SPAM dataset, it was established to be higher than both (Hassan *et al.*, 2022). The work, which has relevance in COVID-19 perception analysis and Industry 4.0, also discusses the challenges of text classification and the necessity of automation for handling large datasets.

This study is dedicated to hierarchical text classification in response to the need for automatization of text accumulation and categorization in terms of the growing amounts of textual information available. This way, it explores the idea of using hierarchical class structures together with other ML models, namely BoosTexter and the Centroid-Boosting algorithm to work on the accuracy of the CIFAR approach and to reduce computing costs (Granitzer, 2023). The study contrasts decisions made by independently trained leaf nodes; it also addresses issues like mistake transmission in the case of hierarchy nodes. The Reuters 21578, Reuters Corpus Volume 1 and Ohsumed datasets were the datasets used during the experiments, and Rocchio as well as support vector machines were employed to do the baseline comparisons. The results also display that the use of hierarchical categorization enhances accuracy and reduces quantitative computing burden. In addition, the suggested model is less erroneous in comparisons between the leaf nodes. The thesis demonstrates the applicability of the automated classification tools for handling large volumes of data pointed by an inherent difficulty of manual categorization which requires domain-specific knowledge and is error-prone. Categorized into principles of traditional indexing, current approaches and problems with hierarchical classification, the thesis gives complete solutions and conclusions.

This study investigates the effectiveness of various text classification methods using data from social media. Ten methods are compared, five are based on lexicon and five are based on ML algorithms using 41 collection of data coming from prominent social media sites (Banik & Agarwal, 2023). Altogether, the study shows that Random Forest and Naive Bayes are the best techniques in capturing human intuition. We also found that for the classification of sentiments into three classes, the best performance was observed in the test using RF while on a small test dataset, NB outperforms other classifiers. As compared with other techniques, the performance of SVMs is generally unremarkable; they are quite often poor performers. Comparing lexicon-based approaches to machine learning-based methods, the Linguistic Inquiry and Word Count method is considerably less efficient. The current study indicates that while techniques such as SVM and LIWC are used more often in marketing, there are more optimal methods of analysis such as NB and RF. It is for this reason that the study presents a discussion on the requirement.

In the study by Hartmann *et al.* (2018), the authors presented a machine learning approach for classification of Arabic text documents based on the embedding and N-grams. It represents documents in a high-dimensional space through trigram frequency vectors. Classification is achieved using the Manhattan distance between documents need to be classed and the training documents with the smallest distance being assigned to the category. The work of this method is compared with Dice similarity coefficient. From the results obtained, the work shows that the measure used namely the Dice coefficient gives better results than the Manhattan distance in the context of the tri-gram text classification. I had to derive classification from the Arabic text, because it is somewhat difficult to work with because it has a different script direction and morphological characteristics from English. Other methods that have been tried before are: distance based, learning, and Bayesian algorithms but the current method of N-grams without stemming exhibits low efficiency. The study relies on newspaper corpus from Jordanian Arabic with training data representing 40% and classification data representing 60% and preprocessing that was done was as follows: elimination of punctuation marks, stop words, anything that is not a letter. The findings reveal that applying the Dice measure improves the classification performance than applying the Manhattan measure in the classification of Arabic texts.

Current trends where we have machine learning systems collecting data from different sources and feeding the data into a reliable knowledge base have been explored (Khreisat, 2009). It evaluates a variety of research journals to determine which algorithm should be deployed and optimized to provide target users with even more personalized and accurate environment. The chapter also has a discussion on the future development of these systems, based on user decision-making and shifting cultural environment. Furthermore, it examines how modifications to the approach to data capture and analysis or the usage of obtained data may open up new avenues for format classifications resulting in more versatile content-based recommendation systems.

The study by Dalal and Zaveri (2011) focuses on Automatic Text Classification of documents. From it, one can learn such things as how to select the appropriate machine learning techniques, retain multiple characteristics, and work with raw text. Indeed, the existence of such difficulties as absent metadata is discussed, along with evaluating statistical and semantic preparation methods. Generally, classification is often associated with text classification for sentiment analysis, spam detection and search results optimization. Specifically, handling blogs, incremental as well as multitopic classification algorithms are required since language is constantly evolving. The study therefore reveals the current existence of the need for complex technologies when it comes to supporting multiple languages and enhancing the performance in areas of text summarization and classification. It also provides information about text classification in general, the approach used, the major issues, and further research potential.

The study by Yao *et al.* (2020) introduces a text classification model that uses the fastText method to overcome challenges of conventional ML techniques, including the curse of dimensionality and low efficiency. Conventional techniques can prove ineffective due to deficiency in working with high-dimensional text representation that does not fully preserve semantics of the texts to support superior classification performance. However, fastText offers a compact dense vector representation of text which provides better performance in classification of text data in contrast to Word2Vec algorithm. The proposed model is used on an emotional polarity judgment of the user comment dataset of Baidu Dianshi platform. As the volumes of electronic text information increase rapidly, the issue of text classification becomes highly significant for extracting desired information and increasing the efficiency of data usage. Based on the fastText model, the paper also shows how the model can deal with large text data by providing certain representations of text to support activities such as differentiation of the emotional polarity of user reviews. It assists the merchants to manage and enhance their products and services while at the same time decreasing the expenses associated with the manual categorization process.

The study by Dharmadhikari *et al.* (2011) discusses several machine learning approaches to text classification on its own, concentrating on supervised, unsupervised and semi-supervised approaches. In terms of importance, it stresses the point that millions of Web documents' storage and searching call for text categorization which plays the key role in information management. Algorithms like Regression models, KNN models, Bayesian techniques, decision trees, Neural Networks, SVMs and many more are discussed in the review. Regarding the structure of the paper, it provides a more detailed analysis of these methods while discussing their usefulness and possibility of applying them. Lastly, the conclusion provides a brief overview of the findings and suggests that more field studies should be conducted.

The study by Gasparetto *et al.* (2022) presents an overview of most current developments in text classification (TC) methods and architectures covering the transition from classical techniques to current deep learning strategies. It emphasizes the differences between the input data and the output labels based on raw text form The transformer-based language models such as BERT and XLNet have been found to provide a very high accuracy on all the datasets. The paper also discusses other approaches like graphs based on where an application of Graph Convolutional Networks (GCNs) is used as well as the hybrid models which combine deep learning and traditional classifier approaches. It compares them on different tasks such as sentiments and questions answering and present new experimental results on multilabel datasets where BERT and XML-R models outperform all other models.

Spam detection is carried out in the study using 12 machine learning classifiers on the dataset of Enron (Occhipinti *et al.*, 2022). This is accompanied by new preprocessing pipeline, which applies various NLP methods for hyperparameters tuning and model improvement. This is because the level of aggression in spam filtering algorithms needs to be balanced according to the study. The effectiveness of the proposed pipeline is that 94% of the F score proves that the suggested pipeline is effective. Moreover, it incorporates the SHAP explain ability methodologies and statistical tests to work on the data and identify significant features which influence the classification. Therefore, to address the ever-rising complexity of spam campaigns as postulated in the paper's conclusion, there is a need to constantly refine the methods of machine learning, NLP, deep learning, etc.

The study by Wu *et al.* (2020) considers and compares deep learning techniques as text classification, including CNNs, RNNs, and Attention Mechanisms with traditional techniques such as Bag-of-Words and TF-IDF, which have significant complications from high dimensionality and sparsity. DL techniques are more suitable for large, high-volume datasets of unstructured or raw data because you do not have to engineer the features. Thus, it is observed that the use of CNNs to capture local attributes, RNNs to deal with sequential input, and Attention Mechanisms to address the problem of long-term dependencies enhances the system's effectiveness.

In this work, machine learning techniques are studied to classify Verbal Autopsy (VA) texts that are interview texts of circumstances of prehospital deaths in underdeveloped nations (Danso *et al.*, 2014). It makes a taxonomy for this purpose and compares different feature value representations, ML algorithms and feature reduction techniques. Through the study, it is found that with respect to all the classifiers, both TF-IDF and normalized word frequency are particularly effective while compared to other methods, such as Support Vector Machines (SVM), and yield significantly higher results. This also indicates their ability to enhance the classification accuracy through a "locally-semi-supervised" feature reduction approach.

In the research by Mitra *et al.* (2007) that sourced data from the Penrose Library at the University of Denver, an architecture based on the Least Square Support Vector Machine (LS-SVM) is used to classify the titles of potentially noisy documents into specified classes. Given a Gaussian Radial Basis Function (GRBF) kernel together with LSI on feature extraction, then the LS-SVM provides an accurate model with higher classification above 99.9% most of the time. Decision confidence levels are set and classification is performed with coefficients calculated by LSI. The paper demonstrates that with the help of LS-SVM, it is possible to design strong learning agents for text classification. Text classifications apply where information is to be sorted, filtered or retrieved in large digital systems such as content management.

In order to determine which method is the most suitable for the summary blog content, the research by Deo and Banik (2022) analyzes three types of text summarization. The challenging task is selecting a sample that best represents the full blog and contained in it and explains it. The purpose of the study is to identify which technique provides the best solution in order to generate precise and reliable summaries of blog postings since blogs are already a significant factor in shaping public opinion and understanding trends in industries with more than 152 million sites.

The work by Kobayashi *et al.* (2017) attempts to show how text classification techniques can be applied in organizational research to filter and sort mass information. In large document collections, the conventional manual classification approaches are not feasible by virtue of reliability, cost and time factors. Apart from giving a detailed guide on how to perform TC involving preparation and classification of the training data, preprocessing, transformation, classification and validation of the data, the paper demonstrates that text mining and machine learning are useful and efficient substitutes to conventional methods. It contains a lesson which demonstrates the job vacancy mining by the authors with the R code and stresses the importance of the model categorization testing. The purpose of the paper in this case therefore is to bring more awareness to organizational scholars on the possibilities of using TC techniques that have been made tractable. Such subjects as political science, law and finance fall under these categories of subjects.

The study by Deng *et al.* (2019) considers feature selection methods for text classification to accommodate heterogeneous multimedia data of text, audio, video and images. It starts with a review of usual approaches to representation of documents and measures of similarity. The study examines and assesses the benefits and drawbacks of four feature selection models, wrapping and inserting, filtering and masking, embedding and hybrid. In addition to these models, it also describes some new feature selection approaches. The review also focuses on how important charged text classification is for processing large amounts of textual information which is extracted from the news and social platforms. This research provides an understanding of the methods employed in the text classification as practiced in the present age.

The effect of current trends of the modern machine learning approaches and word embeddings on the HTC is reviewed here in this work (Stein *et al.*, 2019). Thus, it uses word embeddings like GloVe, word2vec and fastText to evaluate models such as XGBoost, SVM, Keras' CNN and fastText. As for the findings of the study, fastText proves to be efficient on the RCV1 dataset as it reported a lcaF1 score of 0. 893. HTC definitions, text representations and classification algorithms are reviewed, trying to pay special attention on the experimental results and analysis. It also addresses the current discourse about the hierarchy versus flat approach to classification processes.

To improve text classification, the study by Marivate and Sefara (2020) compares and analyzes several text augmentation strategies. In this, it assesses procedures like Word2Vec-based augmentation or the instance generation process, mix-up, round-trip translation, etc., which are discussed in this paper using SSMs and News dataset of articles. Our work demonstrates that Word2Vec-based augmentation is most beneficial when formal synonym models cannot be utilized, while mix-up enhances the model's performance and discourages overfitting. Round-trip translation is less possible because of the expenses involved. It also discusses data augmentation technique and its significance in enhancing the performance and robustness of the model especially when there are few samples of labeled data.

The study by Lodhi *et al.* (2002) focuses on text categorization found by kernel of length κ subsequences. In an effort to emphasize almost continuous sequences, this kernel first decreases exponentially the weight assigned and second calculates the inner product within the feature space that these subsequences form. While the direct computation of this kernel is very intensive in terms of processing, evaluation might be performed quite efficiently due to the use of dynamic programming. The approach is shown to be practically effective for small sets of data and appears to be more effective than some standard techniques such as the word feature space kernel. While using the Gramme matrix in analysis, for bigdata, there is always the ability to apply the approximation technique to optimize the rate of computation in analysis. Unlike some of the other methods that use feature vectors extracted directly from the documents for text categorization like n-gram and bag of words, this method attains superior accuracy by employing sequence alignment.

In the research by Wei *et al.* (2018), the authors proposed the use of deep learning algorithms like CNNs for the assessment of the legal documents based on the predictive coding. It then assesses the proposed CNN against other real legal datasets in comparison to other traditional machine learning methods including LR and SVM. This research shows that CNNs can enhance the text classification in the legal field since CNNs perform better with more massive training datasets.

Focusing on text categorization, the work by Dixon *et al.* (2018) introduces a new methodology for measuring and minimizing the random error in machine learning models. Intentional bias is defined using a test set and particular input attributes such as demographic identifying terms. This technique is used by the authors to analyze and mitigate bias in text classifiers particularly on the level of toxicity in Wikipedia Talk Pages. They found that various forms of bias such as false positive in toxicity detection due to imbalance of certain phrases may occur as a result of such imbalances. This paper provides an approach to improve the balance in the training dataset using an unsupervised approach that complements the removal of bias and does not degrade model performance. Further, the work provides instruments for estimating and minimizing untargeted bias in Machine learning Models.

The paper by Padurariu and Breaban (2019) discusses the issue of training a model for the imbalanced text dataset, specifically, the set of job descriptions in the Human Resources field where the text data is sorted into different job types. This paper presents various text representations, classification methods and techniques of balancing the classes for better classification. The study presents a new CSA-based approach that has been optimized by Differential Evolution algorithm to work with class and instance costs. This method shows that it has better impact on imbalanced datasets, which involves the use of more than one algorithm and techniques that can increase the rate of classification and balance.

In the study by Banerjee *et al.* (2022), the authors focus on the importance of the steps related to text preprocessing in the TC and text mining, and illustrate various preprocessing techniques that can improve classification accuracy and speed. It employs BBC English dataset and classifies using chi-square and TF-IDF then cosine similarity score, and shows that preprocessing does enhance the classification result at lower t values. The paper also points out the fact that preprocessing contributes between 50 to 80% of the total time utilized to classify the papers during the machine learning-based text classification. It offered an illumination of an effect of several preprocessing methods on English text for categorization by machine learning algorithms.

The work by Kadhim (2018) is discussed in the context of enhancing the Human Computer Interaction by easing natural languages using NLP. NLP therefore unleashes the power of machine learning and stops requiring human-supplied rules of language. It will also give an overview of several techniques like NLP techniques with the latest approaches for each, which includes techniques like Sentiment Analysis, Named Entity Recognition, Punctuation Restoration and others. It presents an overview of these recent advances in these NA techniques focusing on their ability to offer more accurate and effective language-based systems for application in numerous areas.

Specifically, the study by Li and Zhan (2018) focuses on evaluating the ability of the Bi-LSTM-CNN model to perform text categorization for long news texts. It proposes a model based on CNN and Bi-LSTM. It is possible to state that the information regarding the word order and context in the CNN is maintained when the Bi-LSTM component fuses together the information educed from the both directions to provide a broader understanding. The research conducted on the THUCNews set of data indicates that the Bi-LSTM-CNN model has a higher login of classification for news text as compared with other benchmark models.

The study by Banik *et al.* (2018) proposes a new semi-supervised approach known as SFE to improve Multinomial Naive Bayes (MNB) for the text categorization task. Unlike other approaches for example Expectation-Maximization (EM) where the performance of MNB may not always be enhanced, SFE enhances the AUC and accuracy of MNB whenever more labeled and or unlabeled samples are included. The SFE uses the term probability estimate individuals from unlabeled data and class conditional probability from labeled data to give comparatively better conditioned log likelihood values than EM. The comparison of SFE and EM on eight text classification datasets demonstrates that SFE achieves better results than the EM in the majority of cases and, at the same time, SFE has less computational complexity, which makes it an optimal solution for the analysis of large datasets.

In the study by Su *et al.* (2011), a kernel-based approach to text categorization on the hierarchical structure is proposed, which does not exclude documents in more than one category. Taking advantage of the theoretical framework of Maximum Margin Markov Network, the classification model depicts the hierarchical relationship as a Markov tree with the edge labeling an exponential family. The authors present a solution for optimization to address the problem that has to run through thousands of training cases and hundreds of the hierarchical classification levels with the help of the incremental conditional gradient ascent in the single-example subspace. The weakness of other classification methods that cannot identify the interrelationship between classes in the context of a hierarchy are overcome by this algorithm. Through the use of the hierarchical structure, it enhances predictive precision through comparison with flat classification models. Numerous studies reveal that this algorithm performs favorably in comparison with other recent studies on hierarchical multicategory or multilabel classification. The paper also explains how one can apply conditional gradient to improve the optimization and to be able to factor in the more complicated loss functions than one, two and three to yield better results in the large-scale hierarchical text classification.

The proposed approach in this research (Rousu *et al.*, 2005) for improving healthcare tweet classification integrates three feature selection techniques sequentially, namely TF-IDF, LSI and BOW. This method enhances the performance of machine learning classifiers: WebKB4 and using disease-related tweet as the two datasets classifiers include SVM, MLP, and RF classifiers. The results demonstrate the importance of integrating the various feature selection techniques for improved categorization of sparse and shorter textual contents such as tweets because they reveal that the formation of features improves categorization.

According to the study by Srivastava *et al.* (2019), text classification involves sorting documents or grouping them into categories by analyzing their contents, and it utilizes machine learning and statistical methods such as KNN, SVM and NB. The process generally follows a hierarchical structure and includes four main phases: text cleaning and preparation of data, feature selection/extraction, training of the classifier and model creation. This underscores the need for automatic classification of texts primarily because of the ever-growing data from the web, which makes it impossible to categorize the data manually.

According to the study by Vasa (2016), WupLeBleu is a newly introduced machine translation (MT) assessment metric with fuzzy logic-based paraphrase matching and n-gram parameters. Unlike other features based on stem matching, WupLeBueld uses the facets of depth, substance and structure of the WordNet taxonomy to evaluate the relation between paraphrases and at the same time consider the essence of paraphrasing by focusing on synonyms and morphological changes in the representation of a sentence. Self-evaluation is more accurate in assessing the most semantically related WupLeBleu which is found to be more correlated with the human judgment in empirical evaluation studies performed on benchmark datasets.

In the study by Lai *et al.* (2015), a model called RCNN for text categorization is described. This model minimizes noise and increases text semantic analysis capacity through recurrent and convolutional features. The results obtained show that the RCNN, in general, can achieve better performance compared to the competitor approaches for various tasks, including writing style classification, sentiment analysis and subject categorization. The model also proves quite efficient in the case of linear temporal complexity in comparison with the text size.

The study by Singh *et al.* (2019) is proficient in document or text categorization, which includes the classification for classifying the sentiment of the news for positive or negative. It explores two Naïve Bayes approaches – Multivariate Bernoulli-NB Classification and Multinomial-NB Classification. The aim is to thereby identify that among these methods which perform well when used for sentiment analysis on the specified dataset. The paper specifically targets text classification in various areas such as spam filtering, health informatics, e-commerce and sentiment analysis of social media.

The recently proposed Bi-Normal Separation (BNS) measure is one of the 12 feature selection methods for text classification considered in this research (Forman, 2003). When constructing the model, the number of samples in the positive class is much larger than the number of samples in the negative class, and BNS is shown to be better than other methods in terms of accuracy rate, F-measure and recall in the evaluation on 229 datasets. Information Gain is known to be accurate. The work reveals how well BNS works on a set of problems and introduces a new method of evaluation that can be helpful for the practitioner in choosing the best feature selection metrics for the datasets.

5.3 CONCLUSION

The conclusion is the consolidation of several studies and research projects, and Table 5.15 consolidates two models, namely SVM and CNN, which leads to a significant enhancement of the classifiers with the BBC News dataset.

As observed, SVM performs well at all times, and it handles complex features in high dimensionality and fine tunes classification boundaries to achieve high recall and precision level. Specifically, the performance of SVM is more satisfying when using features such as Chi-square feature selection. In addition, it has very high precision and reliability in tasks related to news classification and sentiment analysis.

In contrast, the CNN models demonstrate good performance in the best of learning the patterns of the text data through the spatial hierarchies of the features. The CNN's accuracy is generally very high and can go as high as 98.2% and it really shows that it is able to do much more than basic text

Table 5.15 Comparison table

Study	*Model*	*Accuracy (%)*	*F1 Score (%)*
Haryanto et al., 2018	Chi-squared + SVM	95.05	95
Mauni et al., 2020	DNN	—	100
Shirsat et al. 2019	NB	96.46	97.33
Shah et al., 2020	KNN + LR	99	—
Bounabi et al., 2017	Lovin + Vote	97.97	—
Mol & Kumar, 2020	CNN + HAN	98.2	98
Razali et al., 2020	SVM	98	98
Karsana et al., 2023	NB+SVM	97	97
Pradhan et al., 2017	SVM	97.67	—
Dutt & Smalley, 2020	SVM	98.69	—
Chowdhury & Roy, 2022	Attention	96.14	96.14
Dadgar et al., 2016	TF-IDF + SVM	—	97.48
Barman & Chowdhury, 2020	SVM	98.69	—
Abdullahi et al., 2021	RNN	96.18	96.19

classification, therefore making it able to compete fully for text classification tasks that are heavy on pattern recognition.

Altogether, the studies show that the employment of SVM as well as CNN is very useful for text classification of the BBC News dataset. SVM is characterized by high accuracy and optimal boundary, while CNN is characterized by the recognition of patterns and high accuracy. The decision as to which of these models to use may vary depending on the type of classification done with advantages in handling and classifying text data offered both ways.

REFERENCES

Abdullahi, S. S., et al. (2021). Deep Sequence Models for Text Classification Tasks. In *2021 International Conference on Electrical, Communication, and Computer Engineering (ICECCE)* (pp. 1–6). Kuala Lumpur, Malaysia. 10.1109/ICECCE52056.2021.9514261

Agarwal, B., & Mittal, N. (2012). Text Classification Using Machine Learning Methods-A Survey. In *Proceedings of the Second International Conference on Soft Computing for Problem Solving (SocProS 2012)*. New Delhi.

Banerjee, A., Kumar, V., & Banik, D. (2022). Better Qualitative Searching for Effecting the Performance of Machine Translation. In *International Conference on Applied Machine Learning and Data Analytics* (pp. 119–130). Springer Nature Switzerland.

Banik, D., & Agarwal, M. (2023). *An End-to-End Comparison among Contemporary Content-Based Recommendation Methodologies*. CRC Press.

Banik, D., Ekbal, A., & Bhattacharyya, P. (2018). WupLeBleu: The Word-net Based Evaluation Metric for Machine Translation. In *15th International Conference on Natural Language Processing*.

Banik, D., Ekbal, A., & Bhattacharyya, P. (2019). Machine Learning Based Optimized Pruning Approach for Decoding in Statistical Machine Translation. *IEEE Access*, 7, 1736–1751. 10.1109/ACCESS.2018.2883738

Barman, D., & Chowdhury, N. (2020). A Novel Semi-supervised Approach for Text Classification. *International Journal of Information Technology*, 12, 1147–1157.

Bhattacharyya, P., Saikh, T., Banik, D., & Ekbal, A. (2023). Machine Translation Evaluation Metrics for Recognizing Textual Entailment. In *19th International Conference on Computational Linguistics and Intelligent Text Processing*.

Bounabi, M., Moutaouakil, K. E., & Satori, K. (2017). A Comparison of Text Classification Methods using Weighted Terms Selected by Different Stemming Techniques. In *Proceedings of the 2nd International Conference on Big Data*.

Chavan, G. S., Manjare, S., Hegde, P., & Sankhe, A. (2014). A Survey of Various Machine Learning Techniques for Text Classification. *International Journal of Engineering Trends and Technology*, 15(6), 288–292.

Chowdhury, J. S., & Roy, T. (2022). Attention Based Encoder Architecture for Automatic Text. *International Journal of Scientific Research in Engineering and Management (IJSREM)*, 6(4). 10.55041/IJSREM12107

Dadgar, S. M. H., Araghi, M. S., & Farahani, M. M. (2016). A Novel Text Mining Approach based on TF-IDF and Support Vector Machine for News Classification. In *2016 IEEE International Conference on Engineering and Technology (ICETECH)* (pp. 112–116). Coimbatore, India. 10.1109/ICETECH.2016.7569223

Dalal, M. K., & Zaveri, M. (2011). Automatic Text Classification: A Technical Review. *International Journal of Computer Applications*, 28(2), 37–40. 10.5120/3358-4633

Danso, S., Atwell, E., & Johnson, O. (2014). A Comparative Study of Machine Learning Methods for Verbal Autopsy Text Classification. *International Journal of Computer Science Issues*, 10(6), 10.48550/arXiv.1402.4380

Deng, X., Li, Y., Weng, J., & Zhang, J. (2019). Feature Selection for Text Classification: A Review. *Multimedia Tools and Applications*, 78, 3797–3816.

Deo, S., & Banik, D. (2022). Text Summarization using Textrank and Lexrank through Latent Semantic Analysis. In *2022 OITS International Conference on Information Technology (OCIT)* (pp. 113–118). Bhubaneswar, India. 10.1109/OCIT56763.2022.00031

Dharmadhikari, S., Ingle, M., & Kulkarni, P. (2011). Empirical Studies on Machine Learning Based Text Classification Algorithms. *Advanced Computing An International Journal*, 2(6), 161–169, 10.5121/acij.2011.2615

Dixon, L., Li, J., Sorensen, J., Thain, N., & Vasserman, L. (2018). Measuring and Mitigating Unintended Bias in Text Classification. In *AIES '18: Proceedings of the 2018 AAAI/ACM Conference on AI, Ethics, and Society*.

Dutt, A., & Smalley, K. (2020). Text Classification of BBC News Articles and Text Summarization. *International Journal of Advance Research, Ideas and Innovations in Technology*, 6(6), 149–153.

Forman, G. F. (2003). An Extensive Empirical Study of Feature Selection Metrics for Text Classification. *Journal of Machine Learning Research*, 3, 1289–1305.

Gasparetto, A., Marcuzzo, M., Albarelli, A., & Zangari, A. (2022). A Survey On Text Classification Algorithms: From Text to Predictions. *Information* 13(2), 83. 10.3390/info13020083

Granitzer, M. (2023). Hierarchical Text Classification using Methods from Machine Learning, Master's Thesis, Graz University of Technology.

Hartmann, J., Huppertz, J., Schamp, C., & Heitmann, M. (2018). Comparing Automated Text Classification Methods. *International Journal of Research in Marketing*, 36(1), 20–38, 10.1016/j.ijresmar.2018.09.009

Haryanto, A. W., Mawardi, E. K., & Muljono. (2018). Influence of Word Normalization and Chi-Squared Feature Selection on Support Vector Machine (SVM) Text Classification. In *International Seminar on Application for Technology of Information and Communication (iSemantic)*, Semarang, Indonesia.

Hassan, S. U., Ahamed, J., & Ahmad, K. (2022). Analytics of Machine Learning-based Algorithms for Text Classification. *Sustainable Operations and Computers*, 3, 238–248.

Kadhim, A. (2018). An Evaluation of Preprocessing Techniques for Text Classification. *International Journal of Computer Science and Information Security*, 16(6), 22–32.

Kamath, C. N. N., Bukhari, S. S., & Dengel, A. (2018). Comparative Study between Traditional Machine Learning and Deep Learning Approaches for Text Classification. In *ACM Symposium on Document Engineering 2018*, Halifax.

Karsana, M. M. N. L. K. M., & Astuti, W. (2023). *Single-Label and Multi-Label Text Classification using ANN and Comparison with Naïve Bayes and SVM*. Jurnal Media Informatika Budidarma.

Khan, A., Baharudin, B., Lee, L. H., & Khan, K. (2010). A Review of Machine Learning Algorithms for Text-Documents Classification. *Journal of Advances in Information Technology*, 1(1), 4–20, 10.4304/jait.1.1.4-20

Khreisat, L. (2009). A Machine Learning Approach for Arabic Text Classification Using N-gram Frequency Statistics. *Journal of Informetrics*, 3(1), 72–77.

Kobayashi, V. B., Mol, S. T., Berkers, H. A., Kismihók, G., & Hartog, D. N. D. (2017). Text Classification for Organizational Researchers: A Tutorial. *Organizational Research Methods*, 21(3), 766–799.

Kumar, R. R., Reddy, M. B., & Pappula, P. (2019). Text Classification Performance Analysis on Machine Learning. *MATTER International Journal of Science and Technology*, 28(20), 691–697.

Lai, S., Xu, L., Liu, K., & Zhao, J. (2015). Recurrent Convolutional Neural Networks for Text Classification. In *Proceedings of the Twenty-Ninth AAAI Conference on Artificial Intelligence*.

Li, C., Zhan, G., & Li, Z. (2018). News Text Classification Based on Improved Bi-LSTM-CNN. In *2018 9th International Conference on Information Technology in Medicine and Education (ITME)* (pp. 890–893). Hangzhou, China. 10.1109/ITME.2018.00199

Lodhi, H., Saunders, C., Taylor, J. S., Cristianini, N., & Watkins, C. (2002). Text Classification using String Kernels. *Journal of Machine Learning Research*, 2, 419–444.

Luo, X. (2021). Efficient English text classification using selected Machine Learning Techniques. *Alexandria Engineering Journal*, 60(3), 3401–3409.

Marivate, V., & Sefara, T. (2020). Improving Short Text Classification Through Global Augmentation Methods. In *International Cross-Domain Conference for Machine Learning and Knowledge Extraction.*

Mauni, H. Z., Hossain, T., & Rab, R. (2020). Classification of Underrepresented Text Data in an Imbalanced Dataset Using Deep Neural Network. In *2020 IEEE Region 10 Symposium (TENSYMP)*, Dhaka, Bangladesh, 997–1000.

McCallum, A., & Nigam, K. (1998). A Comparison of Event Models for Naive Bayes Text Classification. In *AAAI Conference on Artificial Intelligence.*

Mishu, S. Z., & Rafiuddin, S. M. (2016). Performance Analysis of Supervised Machine Learning Algorithms for Text Classification. In *2016 19th International Conference on Computer and Information Technology (ICCIT)* (pp. 409–413). Dhaka, Bangladesh. 10.1109/ICCITECHN.2016.7860233

Mitra, V., Wang, C.-J., & Banerjee, S. (2007). Text Classification: A Least Square Support Vector Machine Approach. *Applied Soft Computing*, 7(3), 908–914.

Mol, E. A. Nismi, & Kumar, M. B. Santosh. (2020). Study on Impact of RNN, CNN and HAN in Text Classification. In *2020 Advanced Computing and Communication Technologies for High Performance Applications (ACCTHPA)*, Cochin, India, 94–102. 10.1109/ACCTHPA49271.2020.9213231

Occhipinti, A., Rogers, L., & Claudio, A. (2022). A Pipeline and Comparative Study of 12 Machine Learning Models for Text Classification. *Expert Systems with Applications*, 201, 2022.

Padurariu, C., & Breaban, M. E. (2019). Dealing with Data Imbalance in Text Classification. *Procedia Computer Science*, 159, 736–745.

Pradhan, L., Taneja, N. A., Dixit, C., & Suhag, M. (2017). Comparison of Text Classifiers on News Articles. *International Research Journal of Engineering and Technology (IRJET)*, 04(03), 2513–2517.

Razali, A., Daud, S. M., Zin, N. A. M., & Shahidi, F. (2020). Stemming Text-based Web Page Classification using Machine Learning Algorithms: A Comparison. *International Journal of Advanced Computer Science and Applications (IJACSA)*, 11(1), 570–576.

Rousu, J., Saunders, C., Szedmak, S., & Taylor, J. S. (2005). Learning Hierarchical Multi-category Text Classification Models. In *ICML '05: Proceedings of the 22nd International Conference on Machine Learning.*

Shah, K., Patel, H., Sanghvi, D., et al. (2020). A Comparative Analysis of Logistic Regression, Random Forest and KNN Models for the Text Classification. *Augmented Human Research*, 5, 10.1007/s41133-020-00032-0

Shirsat, V. S., Jagdale, R. S., & Deshmukh, S. N. (2019). Sentence Level Sentiment Identification and Calculation from News Articles Using Machine Learning Techniques. In Iyer, B., Nalbalwar, S., & Pathak, N. (Eds.), *Computing, Communication and Signal Processing. Advances in Intelligent Systems and Computing*, vol. 810. Springer, Singapore. 10.1007/978-981-13-1513-8_39

Singh, G., Kumar, B., Gaur, L., & Tyagi, A. (2019). Comparison between Multinomial and Bernoulli Naïve Bayes for Text Classification. In *2019 International Conference on Automation, Computational and Technology Management (ICACTM)* (pp. 593–596). London, UK. 10.1109/ICACTM.2019.8776800

Srivastava, S. K., Singh, S. K., & Suri, J. S. (2019). Effect of Incremental Feature Enrichment on Healthcare Text Classification System: A Machine Learning Paradigm. *Computer Methods and Programs in Biomedicine*, 172, 34–51.

Stein, R. A., Jaques, P. A., & Valiati, J. F. (2019). An Analysis of Hierarchical Text Classification Using Word Embeddings. *Information Sciences*, 471, 216–232.

Su, J., Shirabad, J. S., & Matwin, S. (2011). Large Scale Text Classification Using Semi-supervised Multinomial Naive Bayes. In *ICML'11: Proceedings of the 28th International Conference on Machine Learning.*

Vasa, N. (2016). Text Classification through Statistical and Machine Learning Methods: A Survey. *International Journal of Engineering Development and Research*, 4, 655–658.

Wei, F., Qin, H., Ye, S., & Zhao, H. (2018). Empirical Study of Deep Learning for Text Classification in Legal Document Review. In *2018 IEEE International Conference on Big Data (Big Data)* (pp. 3317–3320). Seattle, WA, USA. 10.1109/BigData.2018.8622157

Wu, H., Liu, Y., & Wang, J. (2020). Review of Text Classification Methods On Deep Learning. *Computers, Materials & Continua*, 63(3), 1309–1321.

Yao, T., Zhai, Z., & Gao, B. (2020). Text Classification Model Based on fast-Text. In *2020 IEEE International Conference on Artificial Intelligence and Information Systems (ICAIIS)* (pp. 154–157). Dalian, China. 10.1109/ICAIIS49377.2020.9194939

Chapter 6

Enhancing performance of a deep neural network utilizing ANOVA-based feature selection for drought prediction

Rajashree Dash, Umesh Kumar Das, and Amiya Maity

6.1 INTRODUCTION

As one of the worst natural disasters, drought has a significant global influence on socio-economic stability and agricultural productivity. Because of its unpredictable nature, areas that depend on a steady supply of water for industry, agriculture and livelihoods face grave risks (Karamouz et al., 2009; Hao et al., 2014; Hao et al., 2018). The frequency and severity of droughts are predicted to increase with the acceleration of climate change, making the problems faced by impacted people more severe. In order to minimize negative effects and prepare for worst-case scenarios, proactive risk management and resource allocation depend heavily on accurate forecasting of drought conditions. Conventional approaches to drought prediction frequently rely primarily on human knowledge and physical labour, both of which have considerable drawbacks despite their value (Fung et al., 2020). These techniques frequently can't handle big amounts of data effectively, can be laborious and are prone to human error. On the other hand, modern predictive analytics applications, especially those that use machine learning (ML) techniques, have shown to perform better when handling multidimensional, complicated information. Among these strategies, ML's Deep Learning subset has become a potent instrument that can identify complex patterns and relationships in data that more conventional approaches would overlook (Bolen et al., 2021; Minixhofer et al., 2021).

Karamouz et al. (2009) have suggested a hybrid drought index (HDI) by integrating three popular drought indices. The authors have used an artificial neural network taking three drought indices in the present month as input to predict HDI for next month. The severity in drought is further examined using this predicted HDI value. Fung et al. (2020) used three variants of support vector regression to forecast month ahead prediction of SPI value to further analyse the severity of agricultural drought from it (Fung et al., 2020). Bolen et al. (2021) has accessed the impact of a Long Short-Term Memory (LSTM) in seasonal drought prediction. The model has shown superior results compared to LR. (Minixhofer et al. (2021) have also

DOI: 10.1201/9781003589716-6

analysed the performance of LSTM for country-level prediction drought condition in the United States. In their study, they have introduced a drought earth dataset including meteorological as well as location identifiers and their corresponding drought scores. Performance of an ELM and a neuro fuzzy system is accessed for drought forecasting using SPI index (Mokhtar et al., 2021; Adnan et al., 2023). A review on ML techniques is presented by Sundararajan et al. (2021). According to their study, most existing articles have used a single classifier, and rare use of hybrid classifiers is found in literature. Less work is done using multiple drought indices. An open-source library for climate modelling and its usage is detailed by Nguyen et al. (2024). Though a number of ML techniques are used for drought prediction, usage of DNN is rare in this field.

To mitigate the research gap, the creation of a simple deep neural network (DNN) for drought prediction across various categories is the main goal of this study. The DNN hopes to deliver precise and trustworthy predictions, acting as an early warning system for efficient water and natural resource management. One of the most important steps in improving the effectiveness of drought prediction predictive models is to identify the most significant features. An optimized subset of relevant features from the original dataset is chosen using the analysis of variance (ANOVA)-based filter feature selection (FS) approach in order to enhance the DNN model's performance. With this approach, top-ranked features are chosen after rating them based on the likelihood of their mean value being identical in different target class. The feature with the lower likelihood value is regarded as more significant to the intended output. After rigorous testing over the drought earth dataset (Minixhofer et al., 2021), which has 10 years' worth of climate data describing 21 unique features, the model's performance is demonstrated in terms of a set of assessment measures, including accuracy, precision, recall and F1 score. The model's performance is also contrasted with a few other conventional ML classifiers, including SVM, DT and LR. The main contributions of the work are as follows:

- A custom five-layer sequential DNN is developed for drought prediction.
- The features that are more significant to the target output are selected for the DNN model utilizing ANOVA.
- The hyper-parameters of DNN are finalized by experimenting with nine different configurations.
- A comparative performance analysis of the custom DNN with three conventional ML classifiers, with and without feature selection, is presented.

The next section presents the working details of the proposed DNN model utilizing ANOVA for drought prediction followed by the findings presented and analysed in Section 6.3 and conclusion in Section 6.4.

6.2 PROPOSED ANOVA-BASED DNN MODEL FOR DROUGHT PREDICTION

This section presents an outline of the computational procedures used to create and improve the proposed simple DNN to forecast various drought categories by utilizing its capacity to identify intricate patterns and relationships in the data. The architecture of the DNN is designed to capture the complex interactions between the input features and the target drought categories. The DNN architecture developed for this task is a sequential model including one input layer, four dense layers each one followed by an associated dropout layer and a final output layer. The first dense layer includes 128 neurons, second dense layer includes 64 neurons, third dense layer includes 32 neurons, and fourth dense layer includes 16 neurons. All the neurons of these dense layers use the Rectified Linear Unit (ReLU), activation function such to introduce non-linearity into the model, enabling it to learn complex patterns. The dropout rates for all the dropout layers are set to 0.5. For classification tasks, the output layer consists of neurons equal to the number of drought categories, with a soft-max activation function to produce probability distributions over the categories. The hyper-parameters such as exact number of dense layers, number of neurons per, learning rate, and batch size are fine-tuned by experimenting with different configurations to achieve optimal performance.

Figure 6.1 shows the workflow diagram of the proposed ML-based drought prediction model. In the data collection stage, the drought earth

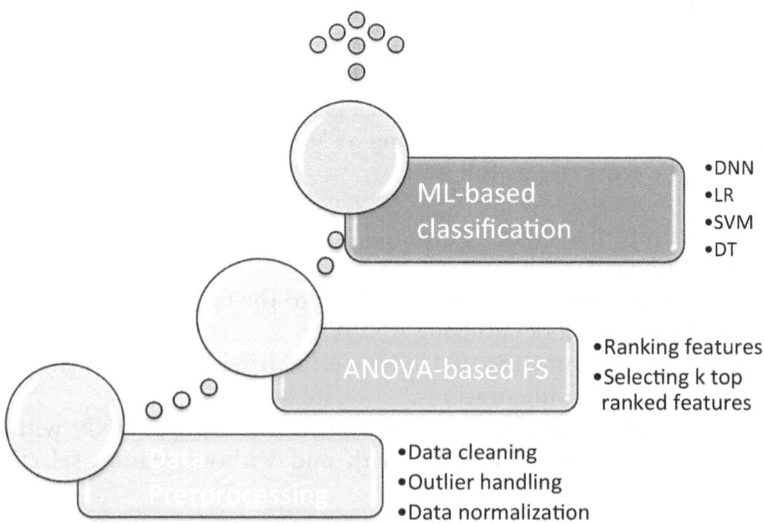

Figure 6.1 Workflow diagram of the proposed ML-based drought prediction model.

dataset is collected from kaggle using the link "https://www.kaggle.com/datasets/cdminix/us-drought-meteorological-data". The dataset consists of globally available meteorological features widely used for drought prediction, paired with location meta-data and the previous drought observations for 20 years. The target variable presented in the dataset represents the weekly drought score, whereas daily values for meteorological data points are available. The drought level score belongs to six categories ranging from 0 to 5 from lowest to highest levels. The score of zero refers to the normal or wet conditions or the natural states of geography. The score of one suggests a condition of abnormally dry situations. A drought score of two refers to moderate droughts and higher dried situations. A drought score of three is referred as severe droughts. Extreme drought conditions are reflected by a drought score of four. At last, a drought score of five is the rarest and is categorized as an exceptional drought situation. After collecting data, in the preprocessing phase the missing values and outliers are handled. In the data cleaning phase, the rows including NAN values are removed from the dataset. The preprocessed dataset includes 1048756 rows and 21 features.

Then ANOVA is utilized in the context of drought prediction to find important characteristics that influence the severity of the drought. With this technique, first the p-value indicating the likelihood of the mean value of the feature being similar in the different target classes is calculated and then the features are ranked based on the p-values. The top-ranked features are selected based on the feature set size provided. The optimal feature set size for the model is decided by simulating over different feature set sizes. ANOVA aids in the selection of the most pertinent features for model training by comparing the means of multiple features across different drought categories. By focusing on important variables and lowering dimensionality, this phase improves model performance and produces predictions that are more accurate and dependable. Then the training and test set are formed with the specified split ratio and the DNN classifier is created. In this study, the DNN classifier is optimized using adam optimizer setting epoch to 100, batch size to 128, sparse_categorical_crossentropy as the loss function. Finally, the trained model's performance is accessed over the test set using four evaluation criteria. Three traditional ML classifiers such as LR, SVM and DT are also trained over the same training set, and their prediction outcomes for the same test set are compared with the outcome of DNN.

6.3 RESULTS AND DISCUSSION

In this study, performance of simple sequential DNN and three traditional ML classifiers utilizing ANOVA-based feature selection technique is observed for drought prediction. The collected drought earth dataset initially contains 2756796 samples with 21 features representing meteorological features, location meta-data and previous weekly drought observations. After handling the missing values and treating the outliers, the sample size reduced to 1048756, which is further scaled using standard_scalar and is divided into training set, test set with a split ratio of 8:2. The performances of the four classifiers are initially observed without using FS as presented in **Table 6.1**. From the observations, it has been clearly found that DNN produces a higher accuracy of 0.94, a precision of 0.91, a recall of 0.94 and F1 score of 0.92, which are better compared to the remaining three classifiers. The epoch-wise accuracy obtained by the DNN model using adam optimizer during its training is shown in Figure 6.2.

Then the performance of the four classifiers is observed using ANOVA-based FS technique. The number of features selected by ANOVA depends on the k value given as its input. The k value of ANOVA specifies the feature set size. To find the optimal k value of ANOVA, the performance of DNN is observed with five different k values. The accuracy obtained by the model using different k values is depicted in Figure 6.3. Analysing the accuracy values with different k values of ANOVA, it is clearly observed that the DNN produces higher accuracy of 0.98 with feature set size 9. With the same feature set size, the performances obtained for the four classifiers are given in **Table 6.2**. From this table, it is clearly observed that all the four evaluation measures have shown improved outcomes for DNN, LR and SVM with FS, whereas for DT, it remains the same. Among all the classifiers, DNN with ANOVA has shown superior results. Further the accuracy of DNN with nine different hyper-parameter configurations is presented in **Table 6.3**. Out of these nine configurations, clearly configuration-1 using five layers with adam optimizer and batch size 128 shows better accuracy compared to remaining ones.

Figure 6.4 illustrates the performance of all the four classifiers for the four criteria using ANOVA-based FS and without using it. Utilizing ANOVA-based FS, DNN has got an improvement of 4.25% in accuracy, 6.59% in precision, 4.25% in recall and 5.43% in F1 score compared to the DNN model without using ANOVA. By comparing the performance of all classifiers with ANOVA, it is clearly observed that DNN has got an accuracy improvement of 27.27%, 10.11% and 78.18% compared to DT, LR and SVM, respectively. Similarly, the precision score of DNN has improved by

Table 6.1 Performance comparison of classifiers (without FS)

Model	*Accuracy*	*Precision*	*Recall*	*F1 Score*
DNN	**0.94**	**0.91**	**0.94**	**0.92**
DT	0.77	0.67	0.77	0.70
LR	0.60	0.41	0.60	0.46
SVM	0.42	0.59	0.41	0.45

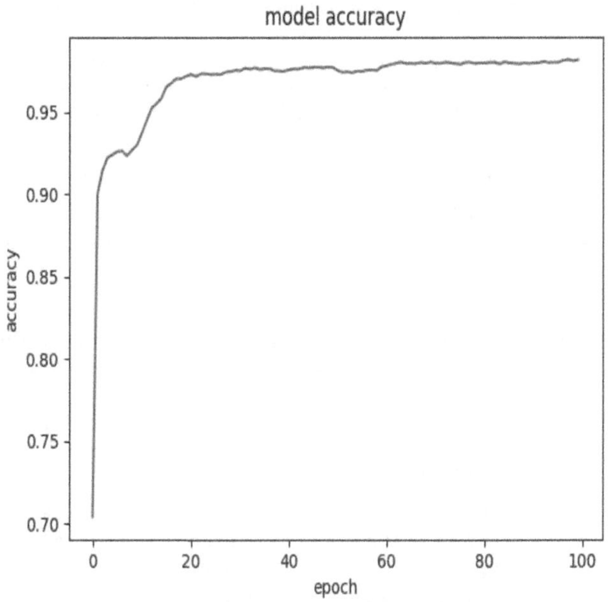

Figure 6.2 Epoch-wise accuracy of DNN during training.

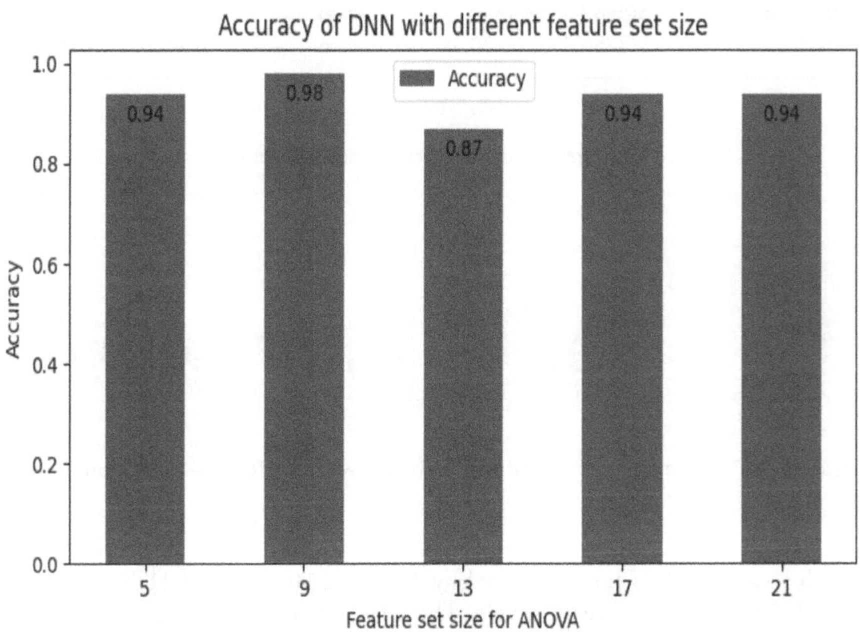

Figure 6.3 Accuracy of DNN with different feature set size.

Table 6.2 Performance comparison of classifiers with ANOVA-based feature selection

Model	*Accuracy*	*Precision*	*Recall*	*F1 Score*
DNN	**0.98**	**0.97**	**0.98**	**0.97**
DT	0.77	0.67	0.77	0.70
LR	0.89	0.86	0.89	0.87
SVM	0.55	0.66	0.55	0.68

Table 6.3 Performance comparison of classifiers with ANOVA-based feature selection

Model configuration	*No. of layers (No of units per layer)*	*Optimizer*	*Batch- size*	*Accuracy*
Model-1	5(128,64,32,16,6)	Adam	128	0.98
Model-2	5(128,64,32,16,6)	Adam	256	0.94
Model-3	5(128,64,32,16,6)	Adam	64	0.94
Model-4	5(128,64,32,16,6)	SGD	128	0.87
Model-5	5(128,64,32,16,6)	Nadam	128	0.94
Model-6	5(128,64,32,16,6)	RMSprop	128	0.87
Model-7	5(96,48,24,12,6)	adam	96	0.94
Model-8	5(486,162,54,18,6)	adam	486	0.98
Model-9	5(3750,750,150,30,6)	adam	3750	0.98

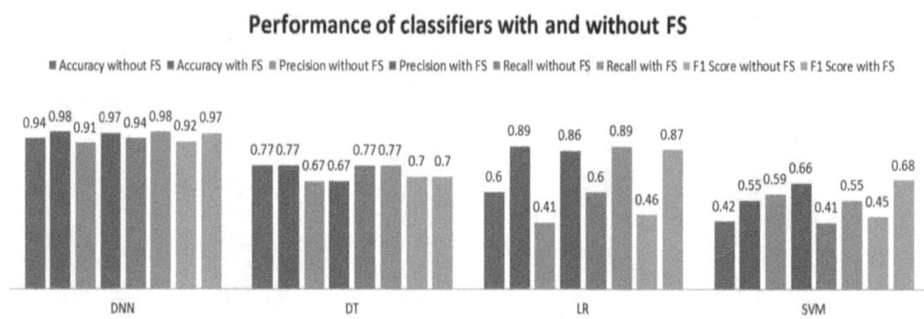

Figure 6.4 Performance of classifiers with and without feature selection.

44.77%, 12.79%, 46.96% compared to DT, LR and SVM, respectively. The recall score of DNN has improved by 27.27%, 10.11%, 78.18% compared to DT, LR and SVM, respectively. The F1 score of DNN has improved by 38.57%, 11.49%, 42.64% compared to DT, LR and SVM, respectively. Overall, DNN with and without ANOVA has shown better results compared to the three traditional ML classifiers.

6.4 CONCLUSIONS

In conclusion, this study successfully demonstrates the potential of a simple sequential five-layer DNN in predicting various categories of drought, utilizing the drought earth dataset collected from kaggle with 21 distinct features. The implementation of ANOVA-based feature selection strategy significantly enhanced the model's performance, as evidenced by the comparative analysis of the classification criteria presented in **Tables 6.1, 6.2** and **Figure 6.4**. The findings underscore the superiority of DNN over traditional ML models in forecasting droughts, offering a reliable and efficient tool for early warning systems. This research not only contributes to the field of predictive analytics but also provides valuable insights for future studies and practical applications in drought management and resource allocation. By leveraging advanced ML methodologies, we can better prepare for and mitigate the adverse effects of droughts on affected regions, thereby ensuring more resilient agricultural practices and sustainable water management.

REFERENCES

Adnan, R. M., Dai, H. L., &Kuriqi, A. (2023). Improving drought modeling based on new heuristic machine learning methods. *Ain Shams Engineering Journal* 14: 102168.

Bolen, B., Sodwatana, M., &Hampson, H. (2021). Drought Prediction in the Western US Using a Long Short-Term Memory (LSTM) Model. *CS230: Deep Learning, Winter 2021, Stanford University, CA. (LateX Template Borrowed from NIPS 2017)*, 1–6.

Fung, K. F., Huang, Y. F., Koo, C. H., & Mirzaei, M. (2020). Improved SVR machine learning models for agricultural drought prediction at downstream of Langat River Basin, Malaysia. *Journal of Water and Climate Change*, *11*(4), 1383–1398.

Hao, Z., AghaKouchak, A., Nakhjiri, N., &Farahmand, A. (2014). Global integrated drought monitoring and prediction system. *Scientific Data*, *1*(1), 1–10.

Hao, Z., Singh, V. P., & Xia, Y. (2018). Seasonal drought prediction: Advances, challenges, and future prospects. *Reviews of Geophysics*, *56*(1), 108–141.

Karamouz, M., Rasouli, K., & Nazif, S. (2009). Development of a hybrid index for drought prediction: case study. *Journal of Hydrologic Engineering*, *14*(6), 617–627.

Minixhofer, C., Swan, M., McMeekin, C., & Andreadis, P. (2021, July). DroughtED: A dataset and methodology for drought forecasting spanning multiple climate zones. In *Tackling Climate Change with Machine Learning: Workshop at ICML 2021*.

Mokhtar, A., Jalali, M., He, H., Al-Ansari, N., Elbeltagi, A., Alsafadi, K., ... & Rodrigo-Comino, J. (2021). Estimation of SPEI meteorological drought using machine learning algorithms. *IEEe Access*, *9*, 65503–65523.

Nguyen, T., Jewik, J., Bansal, H., Sharma, P., & Grover, A. (2024). Climatelearn: Benchmarking machine learning for weather andclimate modeling. *Advances in Neural Information Processing Systems*, 36, pp. 75009–75025.

Sundararajan, K., Garg, L., Srinivasan, K., Bashir, A. K., Kaliappan, J., Ganapathy, G. P., ...& Meena, T. (2021). A contemporary review on drought modeling using machine learning approaches. *Computer Modeling in Engineering & Sciences*, *128*(2), 447–487.

Chapter 7

Protein–protein interaction prediction using extended natural vector representation of proteins and comparison with natural vector method

Shreya Swarnaker and Subhashis Chatterjee

7.1 INTRODUCTION

One or more long chains comprising 20 standard amino acid residues make up the massive biological macromolecules known as proteins. They carry out almost all biological processes and cellular activities in organisms. The majority of the time, nevertheless, they collaborate with other proteins in their environment to complete their jobs rather than performing their functions in isolation. In most of the cellular biological processes in organisms, for example, signal transduction, metabolic pathways, the immune response system, transcription control processes and cellular structure, protein-protein interactions (PPI) are very important. Notably, interactions between complex systems' components appear to be more closely linked to their functions than individual components' features (Pawson, 2004). Hence, PPI analysis is crucial, because it may help with diagnosis of diseases (You et al., 2010), medication target identification, drug discovery, and therapy design (Skrabanek et al., 2008). Many experimental techniques from biochemistry, molecular biology, and genetics (e.g., co-immunoprecipitation, crosslinking, phage display, etc.) have long been employed to discover new PPIs, but due to their inadequateness, these merely assist in covering a minimal portion of the entire database of PPIs (Fields & Song, 1989; Zhang & Wei, 2015). On the other hand, high-efficiency experimental techniques namely yeast two-hybrid screens (Y2H) (Ito et al., 2001), Tandem affinity purification (TAP) (Gavin et al., 2002) as well as mass spectrometric protein complex identification (MS-PCI) (Ho et al., 2002) happened to produce a ton of data, but are costly, wearying and sluggish. Moreover, these techniques might not work with proteins from distinctive organisms and as a result frequently yield false-positive outcomes (Collins et al., 2007). As a result, high-throughput computational approaches are required to accurately and effectively discover PPIs. The precision and caliber of PPI predictions are expected to get enhanced when computational techniques are combined with experimental techniques.

Proteins interact directly through their three-dimensional structures, but because there are so many potential orientations, their interactions are

DOI: 10.1201/9781003589716-7

usually rather limited (Jones & Thornton, 1996). One explanation for this is that the information needed to effectively navigate the biophysical interactions between proteins and fold them into their unique shapes is encoded within the peptide chain. The architectures of amino acids and their electrostatic concentrations are complementary, as demonstrated by the methodical investigation of PPI interactions (Jones & Thornton, 1996; Sheinerman, Norel, & Honig, 2000). The structure and residue arrangement of a protein determines its shape and, in turn, its capacity to link up with another protein. For all proteins, the fundamental sequence of amino acids appears to be the most comprehensive form of information. Therefore, finding and analyzing the properties of those sequences are more convenient when there is an appropriate numerical representation.

Many research efforts rely on preset characteristics retrieved from proteins to address the problem. Examples of these characteristics include conjoint triads (CT) (Shen et al., 2007), autocovariance (AC) (Guo, Yu, Wen, & Li, 2008), ontological features of amino acids (Guo, Yu, Wen, & Li, 2008), composition-transition-distribution (CTD) descriptors (Jansen et al., 2003; Yang et al; 2010), etc. These attributes provide an overview of particular qualities of amino acid sequences, for example, their physicochemical characteristics, distributions in terms of positions, and the occurrences of regional patterns. In addition to these characteristics, a number of statistical learning methods are used to forecast PPIs (Guo, Yu, Wen, & Li, 2008; Huang et al., 2015; You et al., 2014 (December); You et al., 2015). Unfortunately, because the retrieved characteristics employed in these methods are focused on particular aspects of protein profiles, they only cover a small amount of interaction information.

To address these shortcomings, deep learning algorithms were introduced, offering robust capabilities to handle massive amounts of data and acquire features that are beneficial for specific purposes. In current years, deep learning architectures have excelled in various biological fields including determination of variations in genes (Anderson, 2018), categorizations of DNA activities (Quang & Xie, 2016), predicting the RNA-binding location (Zhang et al., 2016) and estimation of the accessibility of chromatin (Min et al., 2017), showing usefulness of convolutional neural networks (CNN) (Anderson, 2018; Zhang et al., 2016) for detecting regional traits autonomously or recurrent neural networks (RNN) (Quang & Xie, 2016) which focus on preserving contextualized information on long-term orders. HNSPPI (Xie et al., 2023) utilizes a feature fusion approach that combines significant knowledge from sequences and topology of networks to achieve extensive acquisition of features. It makes use of a straightforward classifier for prediction, ensuring that it remains lightweight and efficient. AutoPPI (Czibula et al., 2021) adopts a scheme depending on the employment of two autoencoders, one for positive and the other for negative interactions and utilizes three different kinds of neural network configurations for the autoencoders. Jha et al. (Jha, & Saha, 2020) retrieved 3d co-ordinate information from proteins followed by application of ResNet50, AC and CT to extract protein features, which was then passed through LSTM to obtain interaction results. Sun et al. (2017) used Stacked autoencoder (SAE) method combined with AC and CT method to examine PPI prediction and had effective results. In another study by Jha et al., (2022) the graph convolutional network (GCN) and graph Attention network (GAT) were used to forecast PPIs where protein features were learnt by using two pretrained language model SeqVeq (Heinzinger et al., 2019) and ProtBERT (Elnaggar et al., 2021).

Although deep learning has demonstrated strong performance in addressing the problem, it is important to note that deploying a deep learning model comes with several crucial challenges. It requires significant memory and storage, and involves high complexity and substantial time investment, particularly for large-scale problems or when handling high-dimensional biological data. This necessitates the use of specialized hardware such as Graphics Processing Units (GPUs) or distributed computing systems, which can increase costs and complicate deployment. Additionally, training and fine-tuning deep learning models often results in substantial time investment, which can be a limiting factor in practical applications. Also, different biological features or characteristics of a protein require different types of representations to be accurately captured and analyzed. The design process should involve thoughtful consideration of the specific biological aspects that are relevant to the task at hand, ensuring that the chosen representation effectively captures these properties. Basically, which biological characteristics of the protein need to be included in the representation is a task-specific problem (Detlefsen, Hauberg, & Boomsma, 2022).

For the purpose of the present research, the natural vector method (Yu et al., 2013) and the extended natural vector method, which is further derived from the chaos game representation of protein (Sun et al., 2020), were employed. To infer interaction or non-interaction between two proteins, their representational vectors were concatenated. Support Vector Machine (Boser, Guyon, & Vapnik, 1992, (July); Cristianini, & Shawe-Taylor, 2000) and Random Forest (Breiman, 2001) are often practiced machine learning technologies and are frequently used in disciplines like computational biology (Byvatov, & Schneider, 2003; Boulesteix et al., 2012). The proposed protein-protein interaction prediction is a problem of binary classification. The problem of overfitting was attempted to reduce by using a large number of dataset while devising the model. In comparison to recently published PPI prediction models, our method's prediction findings are more desirable. Five sections make up the structure of the chapter. The description of the dataset used to validate the model is described in Section 7.2. The protein representation techniques, the conventional support vector machine, random forest algorithm and our proposed approach are explained in Section 7.3. Section 7.4 displays the findings and analysis of the experiment. At last, Section 7.5 draws conclusions.

7.2 DATASET

The protein-protein interaction datasets of human organisms are adopted in this study. The website http://www.csbio.sjtu.edu.cn/bioinf/LR_PPI/Data.htm provides access to the Pan's human dataset (Pan, Zhang, & Shen, 2010). The protein couples which interact with each other, i.e., the positive pairs contained in the above dataset were gathered from HPRD, 2007 edition. By randomly combining proteins with various subcellular localizations, negative interaction pairs were produced. The Pan's human negative protein-protein interaction dataset consists of protein pairs created applying the procedure outlined above, with negative examples sourced from the Negatome database (Smialowski et al., 2010).

In the aforementioned website, there are five datasets: In dataset A, there are 36,630 protein-protein pairs involving a total of 9476 proteins, indicating positive interactions. In B, there are 36,480 pairs from 2184 proteins denoting the negative interactions. While C and D contain proteins having identity below 25%, C holds 3899 pairs from 2502 proteins denoting positive interaction and D has 4262 pairs from 661 proteins constituting the negative ones. In E, there are 1882 protein pairs involving 842 proteins where the first half of these pairs (50%) represent interacting proteins and the second half represent non-interacting ones.

We considered datasets A, B, C and D for our study and the corresponding outcomes are portrayed in the Results section.

7.3 METHODOLOGY

7.3.1 Natural vector (NV)

This is an efficient, alignment-free technique for numerically representing molecular sequences. Yu et al. (2013) demonstrated how a distance measure derived from the normalized distribution of amino acids can be used to define the protein space in 60-dimensional Euclidean space, which represents the protein universe. Despite their intention to depict the protein universe as a subspace of R^{20N+20}, it was observed that the information obtained via 80-dimensional natural vector mapping was only limited to that of 60. Solely based on the total count of each amino acids and their positions in a protein sequence, the mapping method creates a protein space in which proteins having alike traits cluster together. Consequently, the Euclidean distance within two points in the newly generated protein space can be regarded as the measure of biological relatedness between proteins in real life. Thorough investigation determined that this natural vector mapping is one-to-one. This and the fact that number of dimensions 60 is enough to render all required information concludes that each protein is represented within the aforementioned protein space while pertaining all its innate biological characteristics.

This method describes that each point in the protein space constitutes the following components:

1. The frequencies of the 20 amino acids in a protein chain, indicated by n_k where $k = 20$
2. The mean value, $\mu_k = \frac{T_k}{n_k}$, where $T_k = \sum_{i=1}^{n_k} s[k][i]$ represents the total distance and $s[k][i]$ appears to be the stretch between the initial amino acid of the sequence (considered to be origin) and the ith one.
3. Normalized second-order central moments denoted by $D_2^k = \sum_{i=1}^{n_k} \frac{\left(s[k][i] - \mu_k\right)^2}{n \times n_k}$, with $n = \sum_k n_k$.

Concatenating all these components, we obtain the Natural Vector representation of a polypeptide chain denoted in the form $\left(n_k, \mu_k, D_2^k\right)$, k = {all natural amino acids}.

7.3.2 Extended natural vector (ENV)

7.3.2.1 *Chaos game representation (CGR)*

Sun et al. (2020) showed that exploring a protein is equivalent to exploring its three-dimensional CGR image. A regular dodecahedron is a polygon having 20 vertices, which is the same as the number of natural amino acids in nature. Hence based on polarities, the amino acids are distributed along the vertices of the dodecahedron such that homologous proteins cluster together. The allocation happens in a way which ensures: Amino acids from positive polar group (R, H, K) and negative polar group (D, E) are arranged adjacently, whereas non-polar (A, I, L, M, F, P, W, V) and uncharged polar group (N, C, Q, G, S, T, Y) are placed on neighboring vertices.

Protein's CGR picture constitutes of a sequence of points that are obtained iteratively from:

$$X_n = (1-u) \times \omega_n + u \times X_{n-1}, n \geq 1$$

whereas $X_0 = (1,1,1)$

Every such picture resides within the dodecahedron which is engraved inside sphere B having unit radius and centering at (1, 1, 1). $\omega_n \in \Omega$ signifies the n-th element in the protein's amino acid chain, where Ω = set of all natural amino acids and u is calculated to be $\frac{\sqrt{5}-1}{\sqrt{5}+2\sqrt{3}-1}$. The coordinates of the initial vertices are accumulated from Sun et al. (2020).

For each $\omega \in \Omega$, B_ω is the ball regulated by ω, having radius u and centered at $X_\omega = u \times (1,1,1) + (1-u) \times \omega$. The ball governed by ω_n will contain X_n, the n-th point of the sequence. Balls regulated by different amino acids refrain from coming into contact with one another and are tangent to each other. The balls $B_{\omega_1\omega_2}$, $\omega_1, \omega_2 \in \Omega$, controlled by dipeptides $\omega_1\omega_2$ and centered at $X_{\omega_1\omega_2} = u \times (1-u) \times (\omega_1 - X_0) + X\omega_2$, enjoy similar properties and have radius u^2. Therefore, the sum of points within B_ω equals the number of times a specific amino acid appears in its protein sequence. Similarly, the sum of points within $B_{\omega_1\omega_2}$ equals the frequency of the dipeptide $\omega_1\omega_2$ in an amino acid chain.

7.3.2.2 *Fabricating ENV*

The anticipated Extended Natural Vector is comprised of dipeptides information acquired from a CGR image. The following are needed for constructing the vector:

(a) Given a 3-D CGR picture of a protein, the occurrence of each dipeptide in the chain, i.e., the total count of points within each ball governed by the 400 dipeptides, needs to be calculated.
(b) Based on this frequency distribution, 400 different types of dipeptides are split into a maximum of 16 groups.
(c) Following the formula:

$$k = round\left(\frac{15}{M-m}\left(a_{\omega_1\omega_2} - m\right) + 1\right)$$

where $a_{\omega_1\omega_2}$ is the frequency of the dipeptide $\omega_1\omega_2$, $\omega_1, \omega_2 \in \Omega$

$$M = \max\left\{a_{\omega_1\omega_2} : \omega_1, \omega_2 \in \Omega\right\}$$

$$m = \min\left\{a_{\omega_1\omega_2} : \omega_1, \omega_2 \in \Omega\right\}$$

and round(x) returns the integer number that is closest to x.

The information is then used to obtain components of ENV as follows:

1. First group of components: $f_i = \frac{n_i}{400}$, and n_i renders how many dipeptides there are in group i $(1 \le i \le 16)$.
2. Constituents of the second group: Mean locations of each group $\mu_{1,i}, \mu_{2,i}, \mu_{3,i}, i = 1,\ldots,16$ given by, $\mu_{1,i} = \frac{1}{n_i} \sum_{s \in Group(i)} x_{s,i}$, $\mu_{2,i} = \frac{1}{n_i} \sum_{s \in Group(i)} y_{s,i}$, $\mu_{3,i} = \frac{1}{n_i} \sum_{s \in Group(i)} z_{s,i}$ where $\left(x_{s,i}, y_{s,i}, z_{s,i}\right)$ is the ball's center governed by dipeptides and $\left(\mu_{1,i}, \mu_{2,i}, \mu_{3,i}\right) = (0,0,0)$ when $n_i = 0$.

3. Constituents of the third set: The normalized variances given by, $D_{i,r,s,u} = \sum_{t \in Group(i)} \frac{\left(x_{t,i} - \mu_{1,i}\right)^r \left(y_{t,i} - \mu_{2,i}\right)^s \left(z_{t,i} - \mu_{3,i}\right)^u}{n_i}$, $i = 0,1,\ldots,15$, and r, s, u are positive integers such that $r + s + u = 2$.

Concatenating all these components, the vector notation of a 3-D CGR image is procured which has a dimension 160. The Euclidean distance between ENVs of proteins represents their biological distance. It has been established that each CGR image and the ENV deduced from it can only depict a single protein sequence. The converse is also true.

In the present study, each protein sequence was first passed through NV (or ENV) method and transformed into 60-dimensional (or 160-dimensional vectors). For two proteins which clearly indicate an interaction or non-interaction between them, their representation vectors are concatenated and further treated as a single entity. Thus, any protein pair from positive or negative dataset is represented by a 120-dimensional (or 320-dimensional) vector. The classification problem here reduces to typically a binary one, i.e., the set of labels is simply {–1, 1}. The data was then standardized and passed through both the random forest and support vector machine algorithms and results from both methods were compared.

7.3.3 Support vector machine (SVM)

In high-dimensional spaces, Support Vector Machines (SVMs) are particularly powerful because they can handle a large number of features effectively. The algorithm works by identifying the optimal hyperplane that maximizes the separation margin between different classes. SVM's ability to use various kernel functions allows it to project data into higher dimensions where a linear separation is possible. This adaptability makes SVMs robust for high-dimensional data, ensuring that the decision boundary is optimized for better classification performance. The choice of kernel and tuning of parameter C, i.e., the regularization parameter, are crucial for the performance of the model. While a small C value allows for a wider margin with some misclassifications, a large C aims for fewer misclassifications but with a narrower margin. In this particular problem, the radial basis function (RBF) has been chosen as the kernel. In RBF, gamma is a crucial parameter that determines the influence of an individual training example, with a smaller value of gamma indicating a larger influence. Clearly, the two parameters C and gamma are to be enhanced. An exhaustive grid search strategy was implemented to optimize C within a specific range, and the appropriate value of C was observed to be 1.0. Gamma was set to "scale", which indicates that the value is autonomously adjusted using the formula:

$$\text{gamma} = \frac{1}{\text{number of features} \times \text{variance of the features}}$$

This automatic scaling helps in standardizing the influence range of the training examples relative to the feature space of the dataset.

7.3.4 Random forest (RF)

Whenever the input domain is really vast, random subspace (RS) feature selection can potentially improve classifier diversity. The ensemble predictor called Random Forest consists of a specific number of decision trees, which are randomly selected subsets of the input features. While the "curse of dimensionality" plagues the majority of classification methods, high dimensionality can be used to the RS feature selection method's advantage. A new query sample that has been coded by an input vector is placed into each tree in the forest in order to be classified. In the training of the decision trees, each split selects the best feature from a small subset of the feature space that is randomly chosen. A predicted class is provided by every decision tree. The class with the most votes is proposed as the random forest's predicted class. A subset of n features is randomly selected from the total of N features while constructing a decision tree, and the feature from the subset with the highest information gain is utilized to split each node.

To get the best result from the classifier, the corresponding parameters were adjusted. Using a grid search methodology, the parameter optimization was carried out. An RF model's training typically involves adjusting three parameters: the quantity of trees required to be grown (referred to as n_estimators), highest depth of the tree (max_depth), the bare minimal samples that must be present at a leaf node (min_samples_leaf) and the rule followed for splitting (splitting _criterion). In this study, n_estimators=300, max_depth=500, min_samples_leaf=5 and splitting _criterion=ginni were observed to provide the optimal value concerning the computational time.

7.4 RESULTS AND DISCUSSION

The models were trained on randomly chosen 75% of the datasets and tested over 25% testing data. The standard performance measures which are applied here to assess the quality of the results are – accuracy, F1 score and MCC (Matthew's correlation coefficient). These similarity metrics are computed using the subsequent formulas:

$$\text{Accuracy} = \frac{\text{TN} + \text{TP}}{\text{TN} + \text{TP} + \text{FN} + \text{FP}}$$

$$\text{F1} = \frac{\text{TP}}{\text{TP} + \frac{1}{2}(\text{FP} + \text{FN})}$$

$$\text{MCC} = \frac{\text{TP} \times \text{TN} - \text{FP} \times \text{FN}}{\sqrt{(\text{TP} + \text{FP})(\text{TP} + \text{FN})(\text{TN} + \text{FP})(\text{TN} + \text{FN})}}$$

Table 7.1 Results obtained from different methods

	SVM (%)			*RF (%)*		
	ACC	*F1 Score*	*MCC*	*ACC*	*F1 Score*	*MCC*
NV (AB)	78.33	60	77.90	97.78	95.60	97.60
NV (CD)	68.69	67.90	58.50	69.40	65.15	69.37
ENV (AB)	93.60	82.60	86.40	**98.30**	97.50	98.20
ENV (CD)	88.04	80.20	88.00	96.61	95.47	96.51

Table 7.2 Comparison against recent Innovative methods

Article	*Method*	*Accuracy (%)*	*F1 score (%)*	*MCC (%)*
Xie et al. (2023)	HNSPPI	94.92	94.94	–
Sun et al. (2017)	AC encoding + stacked autoencoder	97.19	–	–
Jha & Saha (2020)	ResNet50 on 3D data, LSTM + Encoder/Decoder on sequence AC/CT embedding	97.77	97.93	98.06
Czibula et al. (2021)	AutoPPI	97.90	97.90	–
Jha & Saha (2020)	GAT (LSTM-based LM)	98.13	**98.73**	95.20
Present Study	ENV + RF	**98.30**	97.50	**98.20**

where TP signifies True Positives; FP signifies False Positives; TN signifies True Negatives and FN signifies False Negative. Table 7.1 displays the results obtained by using support vector machine and random forest on each of the considered dataset AB and CD while using both the protein representation techniques.

For the performance comparison task, some recent benchmark works which have reported training accuracy over human dataset were considered. Table 7.2 presents a detailed overview of the performance comparison, while briefly mentioning the methods they have used and the corresponding performance metrics. The "_" sign has been used to indicate the unavailability of metrics from the corresponding articles.

The bold numbers in Tables 7.1 and 7.2 demonstrate the highest reported results of the three metrics for the models. It shows that our model achieves high precision and MCC score.

Random forest, although being very simple machine learning method, surpasses many other complex deep learning models. Even when proteins have a similarity of ≤25%, the model achieves an accuracy of 96.61%. Clearly, the representation vectors aided much in this work. It is therefore evident that 3D image from CGR contains features which are significant for understanding interactions between proteins. No significant information is lost during the conversion of the CGR image to an ENV, and a protein pair is efficiently portrayed through a 320-D vector. The overall method is significantly simpler and more time-efficient compared to previously known methods, making it more practical and accessible.

7.5 CONCLUSION

In the above study, a fundamental methodology that significantly enhances sequence-based PPI prediction is presented. The protein amino acid sequences are first transformed into numerical representations using Natural Vector and Extended Natural vector representations and then concatenated to portray a unified representation of a protein pair. These representations serve as effective mappings of the protein sequences into a mathematical space, capturing critical information about the sequence composition and its potential interactions, enabling better prediction quality. The training data was then passed through two robust machine learning algorithms: support vector machine and random forest algorithms to train them. After training, the models are evaluated on test data to assess their generalization capability and predictive performance. The evaluation results demonstrate that this methodology not only achieves impressive accuracy but also the model can scale effectively across different data sizes and is pertinent to a wide range of biological problems, while requiring minimal adjustment of the parameters. This characteristic is particularly valuable in biological research, where the underlying data may vary significantly across different studies or organisms. The experimental findings demonstrate that this study surpasses many recent cutting-edge techniques on the aforementioned dataset while significantly reducing running time. This reduction in running time is especially important in practical applications where computational resources are limited, and time efficiency is a priority. It also highlights the fact that building a proper model to describe a biological sequence is as much important. While advanced algorithms and powerful computational techniques are central to achieving high performance, the representation of the biological data itself – how the sequences are encoded and presented to the model – is just as crucial. The study highlights the need for thoughtful consideration of the biological context when designing such models, ensuring that they are both computationally efficient and biologically relevant.

REFERENCES

Anderson, C. (2018). Google's AI tool DeepVariant promises significantly fewer genome errors. *Clinical OMICs*, *5*(1), 33–33.

Boser, B. E., Guyon, I. M., & Vapnik, V. N. (1992, July). A training algorithm for optimal margin classifiers. In *Proceedings of the fifth annual workshop on Computational learning theory* (pp. 144–152).

Boulesteix, A. L., Janitza, S., Kruppa, J., & König, I. R. (2012). Overview of random forest methodology and practical guidance with emphasis on computational biology and bioinformatics. *Wiley Interdisciplinary Reviews: Data Mining and Knowledge Discovery*, 2(6), 493–507.

Breiman, L. (2001). Random forests. *Machine Learning*, *45*, 5–32.

Byvatov, E., & Schneider, G. (2003). Support vector machine applications in bioinformatics. *Applied Bioinformatics*, 2(2), 67–77.

Collins, S. R., Kemmeren, P., Zhao, X. C., Greenblatt, J. F., Spencer, F., Holstege, F. C., ... & Krogan, N. J. (2007). Toward a comprehensive atlas of the physical interactome of *Saccharomyces cerevisiae*. *Molecular & Cellular Proteomics*, 6(3), 439–450.

Cristianini, N., & Shawe-Taylor, J. (2000). *An introduction to support vector machines and other kernel-based learning methods*. Cambridge university press.

Czibula, G., Albu, A. I., Bocicor, M. I., & Chira, C. (2021). AutoPPI: An ensemble of deep autoencoders for protein–protein interaction prediction. *Entropy*, *23*(6), 643.

Detlefsen, N. S., Hauberg, S., & Boomsma, W. (2022). Learning meaningful representations of protein sequences. *Nature Communications*, *13*(1), 1914.

Elnaggar, A., Heinzinger, M., Dallago, C., Rehawi, G., Wang, Y., Jones, L., ... & Rost, B. (2021). Prottrans: Toward understanding the language of life through self-supervised learning. *IEEE Transactions on Pattern Analysis and Machine Intelligence*, *44*(10), 7112–7127.

Fields, S., & Song, O. K. (1989). A novel genetic system to detect protein–protein interactions. *Nature*, *340*(6230), 245–246.

Gavin, A. C., Bösche, M., Krause, R., Grandi, P., Marzioch, M., Bauer, A., ... & Superti-Furga, G. (2002). Functional organization of the yeast proteome by systematic analysis of protein complexes. *Nature*, 415(6868), 141–147.

Guo, Y., Yu, L., Wen, Z., & Li, M. (2008). Using support vector machine combined with auto covariance to predict protein–protein interactions from protein sequences. *Nucleic Acids Research*, *36*(9), 3025–3030.

Heinzinger, M., Elnaggar, A., Wang, Y., Dallago, C., Nechaev, D., Matthes, F., & Rost, B. (2019). Modeling aspects of the language of life through transfer-learning protein sequences. *BMC Bioinformatics*, *20*, 1–17.

Ho, Y., Gruhler, A., Heilbut, A., Bader, G. D., Moore, L., Adams, S. L., ... & Tyers, M. (2002). Systematic identification of protein complexes in *Saccharomyces cerevisiae* by mass spectrometry. *Nature*, 415(6868), 180–183.

Huang, Y. A., You, Z. H., Gao, X., Wong, L., & Wang, L. (2015). Using weighted sparse representation model combined with discrete cosine transformation to predict protein-protein interactions from protein sequence. *BioMed Research International*, *2015*(1), 902198.

Ito, T., Chiba, T., Ozawa, R., Yoshida, M., Hattori, M., & Sakaki, Y. (2001). A comprehensive two-hybrid analysis to explore the yeast protein interactome. *Proceedings of the National Academy of Sciences*, 98(8), 4569–4574.

Jansen, R., Yu, H., Greenbaum, D., Kluger, Y., Krogan, N. J., Chung, S., ... & Gerstein, M. (2003). A Bayesian networks approach for predicting protein-protein interactions from genomic data. *Science*, *302*(5644), 449–453.

Jha, K., & Saha, S. (2020). Amalgamation of 3d structure and sequence information for protein–protein interaction prediction. *Scientific Reports*, *10*(1), 19171.

Jha, K., Saha, S., & Singh, H. (2022). Prediction of protein–protein interaction using graph neural networks. *Scientific Reports*, *12*(1), 8360

Jones, S., & Thornton, J. M. (1996). Principles of protein-protein interactions. *Proceedings of the National Academy of Sciences*, *93*(1), 13–20.

Min, X., Zeng, W., Chen, N., Chen, T., & Jiang, R. (2017). Chromatin accessibility prediction via convolutional long short-term memory networks with k-mer embedding. *Bioinformatics*, *33*(14), i92–i101.

Pan, X. Y., Zhang, Y. N., & Shen, H. B. (2010). Large-Scale prediction of human protein– protein interactions from amino acid sequence based on latent topic features. *Journal of Proteome Research*, *9*(10), 4992–5001.

Pawson, T. (2004). Specificity in signal transduction: from phosphotyrosine-SH2 domain interactions to complex cellular systems. *Cell*, *116*(2), 191–203.

Quang, D., & Xie, X. (2016). DanQ: a hybrid convolutional and recurrent deep neural network for quantifying the function of DNA sequences. *Nucleic Acids Research*, *44*(11), e107–e107.

Sheinerman, F. B., Norel, R., & Honig, B. (2000). Electrostatic aspects of protein–protein interactions. *Current Opinion in Structural Biology*, *10*(2), 153–159.

Shen, J., Zhang, J., Luo, X., Zhu, W., Yu, K., Chen, K., ... & Jiang, H. (2007). Predicting protein–protein interactions based only on sequences information. *Proceedings of the National Academy of Sciences*, *104*(11), 4337–4341.

Skrabanek, L., Saini, H. K., Bader, G. D., & Enright, A. J. (2008). Computational prediction of protein–protein interactions. *Molecular Biotechnology*, *38*, 1–17.

Smialowski, P., Pagel, P., Wong, P., Brauner, B., Dunger, I., Fobo, G., ... & Ruepp, A. (2010). The Negatome database: a reference set of non-interacting protein pairs. *Nucleic Acids Research*, *38*(suppl_1), D540–D544.

Sun, T., Zhou, B., Lai, L., & Pei, J. (2017). Sequence-based prediction of protein protein interaction using a deep-learning algorithm. *BMC Bioinformatics*, *18*, 1–8.

Sun, Z., Pei, S., He, R. L., & Yau, S. S. T. (2020). A novel numerical representation for proteins: three-dimensional chaos game representation and its extended natural vector. *Computational and Structural Biotechnology Journal*, *18*, 1904–1913.

Xie, S., Xie, X., Zhao, X., Liu, F., Wang, Y., Ping, J., & Ji, Z. (2023). HNSPPI: a hybrid computational model combing network and sequence information for predicting protein–protein interaction. *Briefings in Bioinformatics*, *24*(5), bbad261.

Yang, L., Xia, J. F., & Gui, J. (2010). Prediction of protein-protein interactions from protein sequence using local descriptors. *Protein and Peptide Letters*, *17*(9), 1085–1090.

You, Z. H., Chan, K. C., & Hu, P. (2015). Predicting protein-protein interactions from primary protein sequences using a novel multi-scale local feature representation scheme and the random forest. *PLoS One*, *10*(5), e0125811.

You, Z. H., Lei, Y. K., Gui, J., Huang, D. S., & Zhou, X. (2010). Using manifold embedding for assessing and predicting protein interactions from high-throughput experimental data. *Bioinformatics*, *26*(21), 2744–2751.

You, Z. H., Zhu, L., Zheng, C. H., Yu, H. J., Deng, S. P., & Ji, Z. (2014, December). Prediction of protein-protein interactions from amino acid sequences using a novel multi-scale continuous and discontinuous feature set. In *BMC bioinformatics* (Vol. 15, pp. 1–9). BioMed Central.

Yu, C., Deng, M., Cheng, S. Y., Yau, S. C., He, R. L., & Yau, S. S. T. (2013). Protein space: a natural method for realizing the nature of protein universe. *Journal of Theoretical Biology*, *318*, 197–204.

Zhang, S. W., & Wei, Z. G. (2015). Some remarks on prediction of protein-protein interaction with machine learning. *Medicinal Chemistry*, 11(3), 254–264.

Zhang, S., Zhou, J., Hu, H., Gong, H., Chen, L., Cheng, C., & Zeng, J. (2016). A deep learning framework for modeling structural features of RNA-binding protein targets. *Nucleic Acids Research*, *44*(4), e32–e32.

Chapter 8

Utilizing an optimized CHPNN model for COVID-19 cases prediction

Smita Mohanty and Rajashree Dash

8.1 INTRODUCTION

A new corona virus (nCov) infection in humans was first identified in the Chinese city of Wuhan in December 2019 (WHO 2020a). This virus causes respiratory diseases in humans starting from common cold to severe acute respiratory disorder. The corona virus, also called as SARS-CoV2 (Severe Acute Respiratory Syndrome Coronavirus 2), has been officially termed COVID-19 by the World Health Organization (WHO 2020b). It has a diversified effect in human body, which may include symptoms like fever, fatigue, dry cough, labored breathing, loss of smell and taste, skin rash, etc. and in comparatively serious cases may lead to multiple organ failure and death as well (WHO 2020c). This disease primarily spreads through person-to-person physical contact, respiratory droplets released by sneezing, cough or by touching surfaces contaminated with these droplets (WHO 2020d). In no time, the Wuhan virus infiltrated many other countries affecting millions of people and brought daily life to a grinding halt worldwide. The exponential increase in corona virus-infected cases across the globe caused the WHO to announce the outbreak a public health crisis of global concern on 30 January 2020 (WHO 2020e). In an attempt to avoid the quick and extensive outspread of the nCov via humans, several countries imposed full or partial shutdowns and lockdowns to break the carrying chain of the virus via humans. Few common preventive measures employed are the use of face mask, washing hands frequently, social distancing, maintaining person hygiene, etc. The very first instance of the COVID-19 case in India was brought forth on 30 January, 2020 in Kerala state. As of 9 December 2021, the number of cases of the COVID-19 disease in India was 268,257,568 and the number of death cases was 5,298,511 according to the *Worldometer* data. The pandemic's profound effects have challenged modern society and completely changed the way we live.

 DOI: 10.1201/9781003589716-8

Keeping in mind the vulnerability of the current situation, the motivation is to come up with an efficient predictive mechanism for COVID. As prevention is better than cure, several predictor models have already been proposed by researchers to forecast its future impact which will be helpful to prepare the potential course of action and decision-making. In the study by Rath et al. (2020), Linear Regression (LR) and Multiple Linear Regression (MLR) models have been proposed to foretell the daily active cases in India as well as in Odisha where MLR has been found to be a strong predictor model. LR model has also been compared with Least Absolute Shrinkage and Selection Operator, Exponential Smoothing and Support Vector Machine techniques (Rustam et al., 2020) to predict the count of confirmed cases, recoveries and death cases in COVID-19 in which Exponential Smoothing outperforms other models. Another study including LR model, Multilayer Perceptron (MLP) and Vector Auto Regression (VAR) has been suggested by Sujath et al. (2020) where MLP gives better prediction results than VAR and LR models. Gene Expression Programming (GEP)-based models also were able to foresee the future behavior of COVID-19 virus in a work introduced by (Salgotra et al., 2020).

In this study, a very well-known bio-inspired algorithm, FPA, has been exploited to optimize the weight parameters during the training phase of the architecturally simple CHPNN model to come up with a robust FPA-based CHPNN model for prediction of future impact of the COVID-19 disease. CHPNN is a very simple single layer neural network with no hidden layers and built around a set of orthogonal Chebyshev polynomials Dehuri & Cho (2010); Mishra & Dash (2014); Dash, (2018) . It can also be called as Chebyshev polynomial-based FLANN or CHFLANN. The absence of hidden layers leads to reduced complexity in computation and hence improved performance in terms of improved forecasting accuracy than MLP as well as Functional Link Artificial Neural Network (FLANN) Mohanty & Dash (2021).. FPA is a very adaptable, flexible and powerful swarm-based optimization algorithm based on the fertilization process of flowering plants Yang, (2012). It has got immense popularity due to its widespread applicability in various fields Mohanty & Dash (2022a); Mohanty & Dash (2022b); Mohanty & Dash (2022c); Mohanty & Dash (2023). In the current work, the weight parameters of the CHPNN model have been optimally tuned by the FPA for forecasting a framework for COVID-19 cases prediction in Odisha and India. The empirical validation of the proposed CHPNN-FPA system is presented with a detailed comparative analysis of other two optimization algorithms like Particle Swarm Optimization (PSO) and Differential Evolution (DE) on six data sets related to corona virus cases.

The next section focuses on the proposed CHPNN-FPA-based COVID-19 predictor model followed by the results and discussion section in Section 8.3. The final observations are presented in Section 8.4.

8.2 PROPOSED CHPNN-FPA-BASED COVID-19 PREDICTOR MODEL

Figure 8.1 represents the overall working methodology of the proposed CHPNN-FPA-based COVID-19 predictor model. The very first step includes collection of COVID-related data set from https://www.covid19india.org/ for daily confirmed, recovered and fatality cases due to COVID-19 pandemic over a time period ranging from 1 July 2020 to 31 October 2021 in Odisha as well as in India. Once data collection is over, data normalization is performed using min-max normalization method and the processed normalized data is partitioned into input and output data by considering the prediction scope and window length. These processes constitute the data preprocessing step. Under data splitting, the preprocessed data is divided into in-sample and out-sample data set in a 70:30 ratio used for training and evaluating the proposed model, respectively. Keeping the expansion order value of CHPNN fixed, the network structure is created by broadening the training data to a higher dimensional data set and initializing a weight vector or population vector with random values in the range of −1 and 1 according to a given population size. Training of the model involves optimizing the weight-related CHPNN parameters over the in-sample data with the help of an optimization technique. Here the learning algorithm used is FPA.

The randomly initialized population vector applied upon the functionally expanded data pattern is then propagated to the hyperbolic tangent activation function, tanh, to generate the output vector. The fitness of individual input with reference to Root Mean Squared Error (RMSE) is evaluated and subsequently optimized by the FPA during the training stage of the network. The process of fertilization in flowering plants serves as the foundation for the development of FPA. This algorithm exhibits two complementary

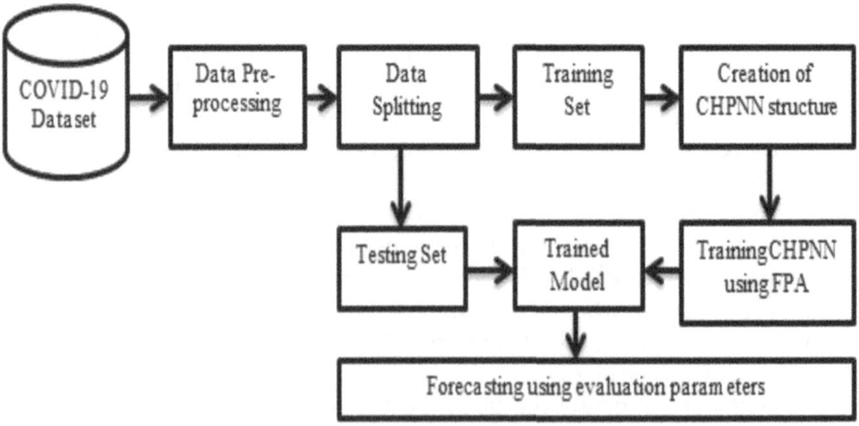

Figure 8.1 CHPNN-FPA-based COVID-19 predictor model.

mechanisms of searching namely global pollination and local pollination. Global pollination process is based on Levy flights and hence biotic in nature. Local pollination is based on local pollen transfer through abiotic agents and hence abiotic in nature. The switching between these two processes is controlled by a random valued parameter named switch probability, p ϵ [0, 1]. The global and local search processes are defined as follows:

$$z_i^{t+1} = z_i^t + \gamma L(\lambda)\left(g^* - z_i^t\right) \tag{8.1}$$

$$z_i^{t+1} = z_i^t + \epsilon\left(z_j^t - z_k^t\right) \tag{8.2}$$

where z_i^t represents the pollen vector in iteration t, γ is the scaling factor that controls the Lévy flight-based step size, $L(\lambda)$ which illustrates the reliability of the reproductive procedure, g^* is the current best population vector in the present iteration. In Eq. (8.2), z_j^t and z_k^t are the pollen vectors derived from various flowers of the same plant in a confined region achieving self-pollination procedure. Mathematically, the expression for self-pollination can be stated as a local random walk with ϵ indicating the [0, 1] range random number that operates as the floral constancy. This control parameter makes sure that adjacent flowering plants are selected more often for fertilization encouraging exploitation over exploration via global pollination. So, parameter, p, has an important role here that is used to switch between these two commonly occurring and expensive operations, respectively. Given that the entire procedure is slightly inclined in favor of local pollination, a value of $p = 0.8$ is utilized for maximum implementations and it is found to perform better. At first g^* is chosen in the initial population with the least RMSE value. Then for a fixed number of iterations, the population matrix or weight matrix undergoes either self-pollination or cross-pollination process depending on the value of randomly chosen p and the fitter one in terms of less RMSE value is updated as the better one in the population matrix. Once all the population vectors go through the process, the g^* value is evaluated in comparison to the current best solution and is updated if the former is better than the later. These steps of fitness value evaluation and g^* value modification are continued till the termination condition is not satisfied. Finally, g^* represents the final and fixed weights for CHPNN, and the model can be used for testing purpose. In order to verify the performance of the proposed predictor model over out-sample data, RMSE is used as the evaluation criteria defined as follows:

$$\text{RMSE} = \sqrt{\frac{1}{N}\sum_{i=1}^{N}\left(s - \hat{s}\right)^2} \tag{8.3}$$

where N stands for the overall set of data, s is the provided data and $\hat{s}$ is the anticipated data obtained from the tested models.

8.3 RESULTS AND DISCUSSION

This study throws light on the improvement in performance of the CHPNN-based COVID-19 predictor framework using FPA as the learning algorithm. Here a total of six data sets are considered referring to daily confirmed, recovered and deceased cases due to COVID-19 pandemic over a time period ranging from 1 July 2020 to 31 October 2021 in Odisha as well as in India. The aim here is to make one day ahead prediction of these three cases over a window size of five days. After setting up the window length and prediction scope to 5 and 1, respectively, the expansion order of CHPNN model is fixed to 2 to create the suitable network structure. This network is trained with three different state-of-the-art learning algorithms named FPA, DE and PSO and then validated upon the test data. For all these implementations, the size of the population is fixed to 20 and the iteration number is fixed to 200. Figures 8.2 and 8.3 depict prediction related to Odisha and India, respectively, in terms of RMSE. In Odisha for the confirmed cases, the proposed CHPNN-FPA model gives a RMSE value of 0.0185, for recovered cases, it gives a RMSE value of 0.0307 and for deceased cases, the model gives a RMSE value of 0.1024, which are less than the RMSE values obtained

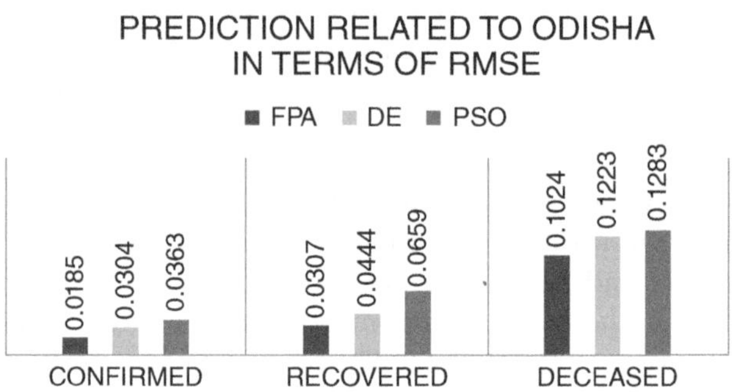

Figure 8.2 Prediction outcome of Odisha using different models.

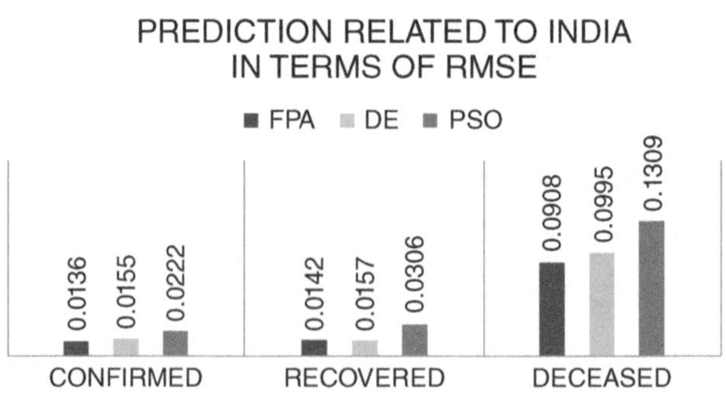

Figure 8.3 Prediction outcome of India using different models.

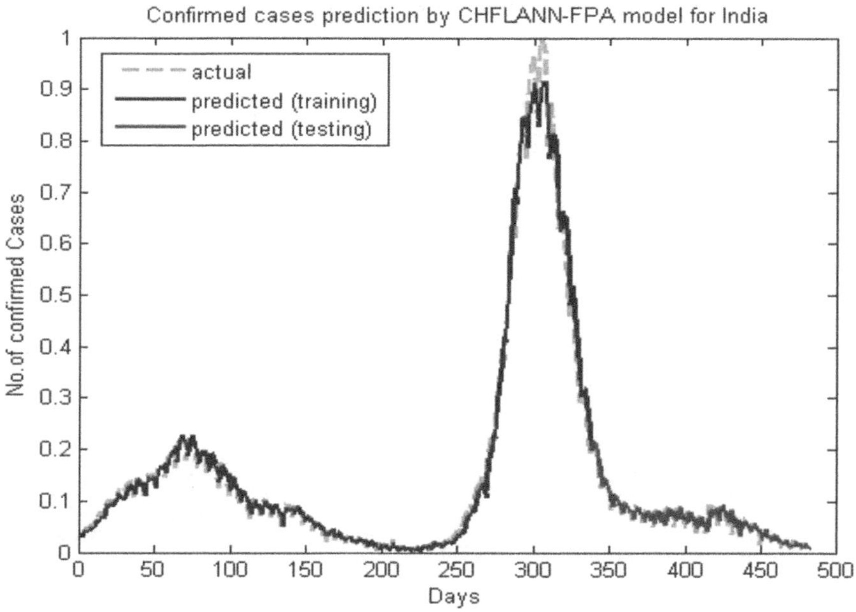

Figure 8.4 Observation of actual vs. forecasted confirmed cases in India using CHPNN-FPA model.

using the CHPNN-DE and CHPNN-PSO models. Similarly, in India also, the proposed CHPNN-FPA model gives a less RMSE value in comparison to the RMSE values obtained using DE and PSO-based CHPNN models. The comparison between actual and predicted number of daily confirmed cases, deceased cases and recovery cases during training as well as testing stages by the suggested model are presented in Figures 8.4–8.9 for India and Odisha, respectively. All the empirical results clearly demonstrate the potential of the CHPNN-FPA model in predicting various types of COVID cases. Also, its superiority over the other two models for the six data sets can be observed in this study.

8.4 CONCLUSIONS

Development of an efficient predictor model to foresee the future impact of corona virus is definitely the need of the hour to make adequate action plan. This chapter has suggested an efficient CHPNN-based COVID predictor model using FPA as the learning technique. A detailed comparative evaluation of the proposed model with CHPNN-PSO and CHPNN-DE models evidently shows that CHPNN-FPA model performs much better in both training and testing phase for all the six data sets with less RMSE value. In future, this work could be further continued with enhanced performance by

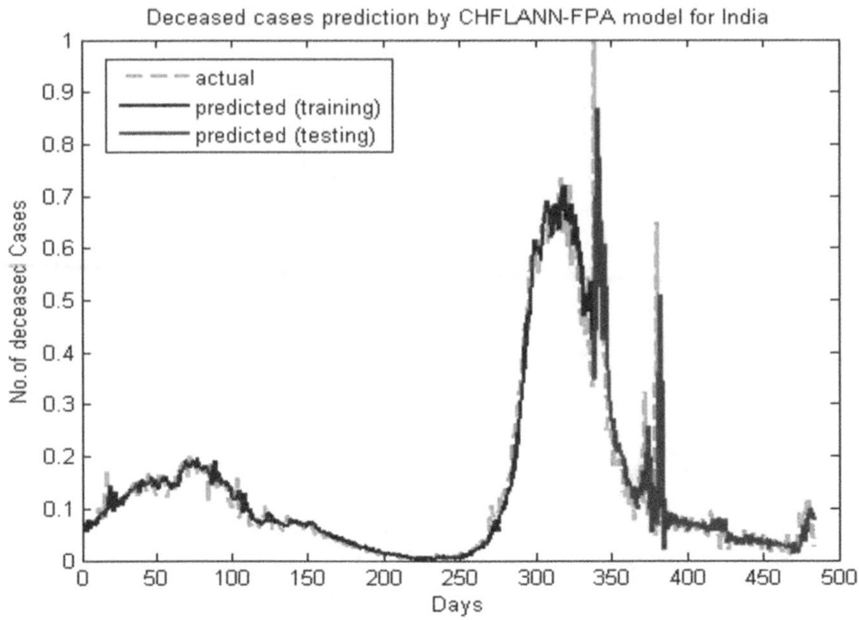

Figure 8.5 Observation of actual vs. forecasted deceased cases in India using CHPNN-FPA model.

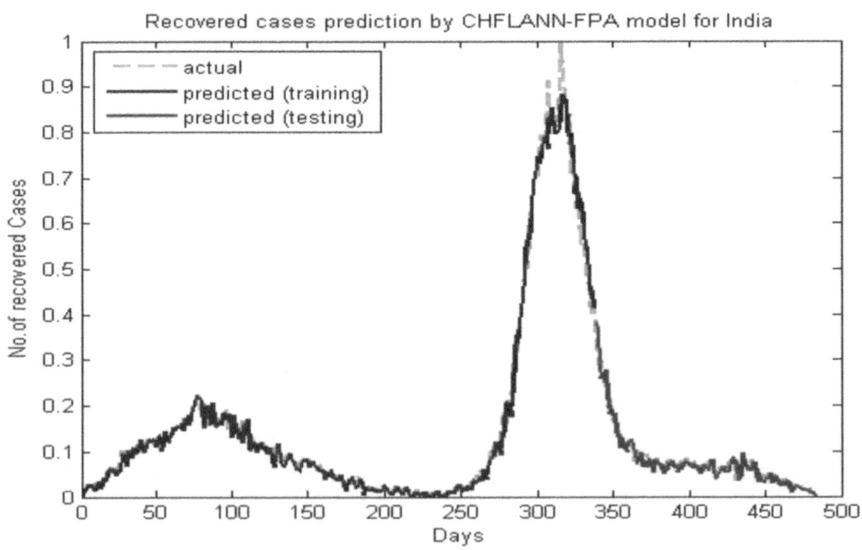

Figure 8.6 Observation of actual vs. forecasted recovery cases in India using CHPNN-FPA model.

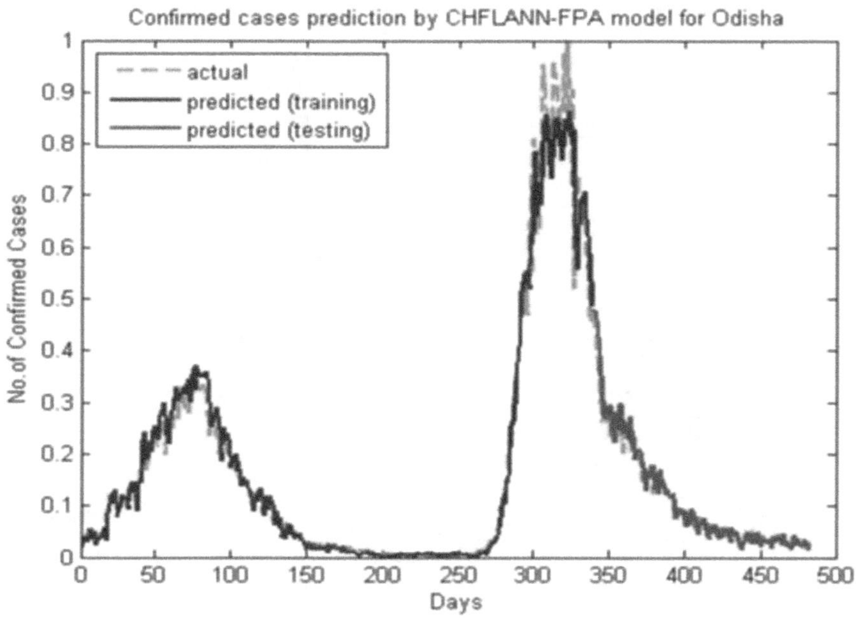

Figure 8.7 Observation of actual vs. forecasted confirmed cases in Odisha using CHPNN-FPA model.

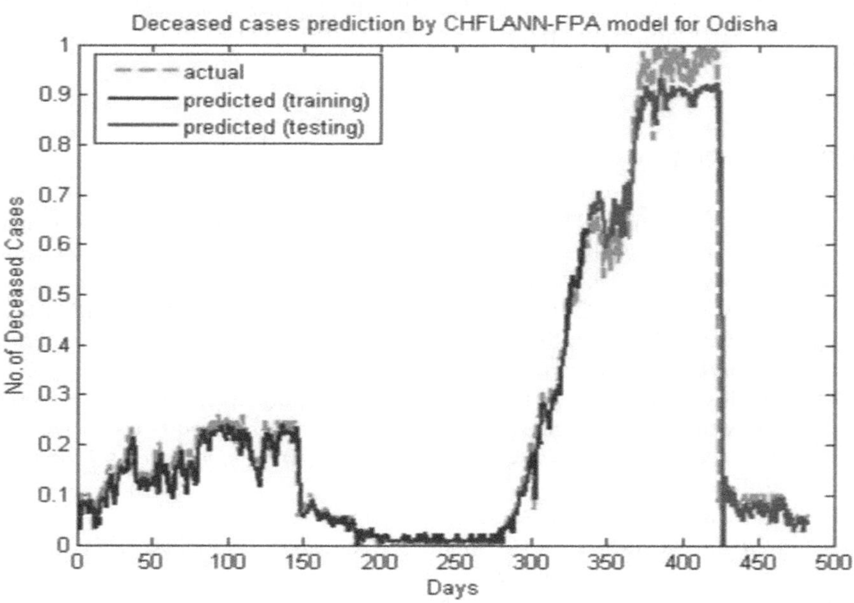

Figure 8.8 Observation of actual vs. forecasted deceased cases in Odisha using CHPNN-FPA model.

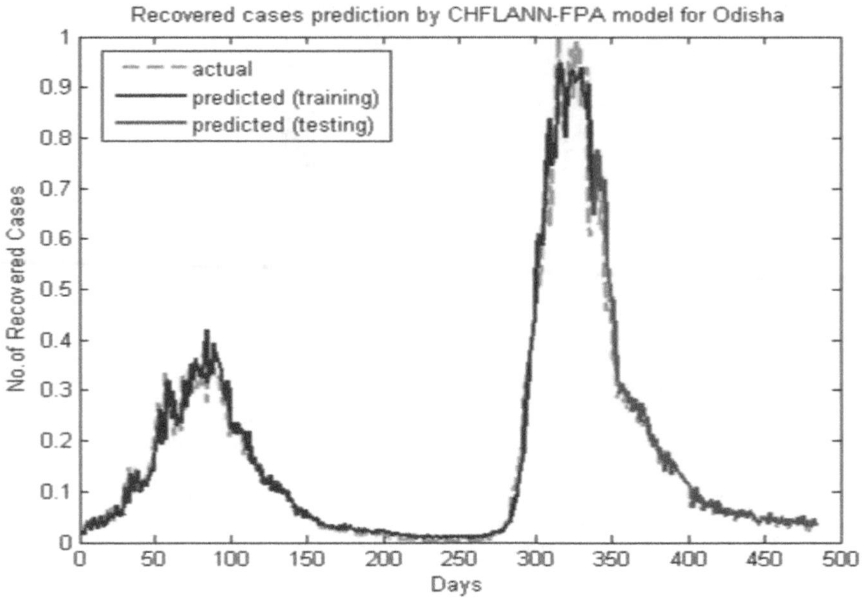

Figure 8.9 Observation of actual vs. forecasted recovery cases in Odisha using CHPNN-FPA model

hybridizing FPA with other learning algorithms and implementing it with different order values and parameter optimization phases of CHPNN-based COVID predictor model.

REFERENCES

Dash, R. (2018). DECPNN: A hybrid stock predictor model using differential evolution and chebyshev polynomial neural network. *Intelligent Decision Technologies*, .12, 93–104.

Dehuri, S., & Cho, S. B. (2010). A comprehensive survey on functional link neural networks and an adaptive PSO–BP learning for CFLNN. *Neural Computing and Applications*, 19, 187–205.

Mishra, M.K., & Dash, R. (2014). A Comparative Study of Chebyshev Functional Link Artificial Neural Network, Multi-layer Perceptron and Decision Tree for Credit Card Fraud Detection. In *2014 International Conference on Information Technology*, 228–233.

Mohanty, S., & Dash, R. (2021). Predicting the price of gold: a CSPNN-DE model. In *Intelligent and Cloud Computing: Proceedings of ICICC 2019*, vol. 2, 289–297. Springer Singapore.

Mohanty, S., & Dash, R. (2022a). A comprehensive review on bio-inspired flower pollination algorithm. *Journal of Information and Optimization Sciences*, 43(5), 963–971.

Mohanty, S., & Dash, R. (2022b). A novel chaotic flower pollination algorithm for modelling an optimized low-complexity neural network-based NAV predictor model. *Progress in Artificial Intelligence*, 11(4), 349–366.

Mohanty, S., & Dash, R. (2022c). Evaluating chaotic functions with flower pollination algorithm for modelling an optimized low complexity neural network based NAV predictor model. *Soft Computing*, 26(18), 9395–9417.

Mohanty, S., & Dash, R. (2023). A flower pollination algorithm based Chebyshev polynomial neural network for net asset value prediction. *Evolutionary Intelligence*, 16(1), 115–131.

Rath, S., Tripathy, A., & Tripathy, A. R. (2020). Prediction of new active cases of coronavirus disease (COVID-19) pandemic using multiple linear regression model. *Diabetes & Metabolic Syndrome: Clinical Research & Reviews*, 14(5), 1467–1474.

Rustam, F., Reshi, A. A., Mehmood, A., Ullah, S., On, B. W., Aslam, W., & Choi, G. S. (2020). COVID-19 future forecasting using supervised machine learning models. *IEEE Access*, 8, 101489–101499.

Salgotra, R., Gandomi, M., & Gandomi, A. H. (2020). Time series analysis and forecast of the COVID-19 pandemic in India using genetic programming. *Chaos, Solitons & Fractals*, 138, 109945.

Sujath, R., Chatterjee, J. M., & Hassanien, A. E. (2020). A machine learning forecasting model for COVID-19 pandemic in India. *Stochastic Environmental Research and Risk Assessment*, 34, 959–972.

WHO. (2020a). Available online at: https://www.who.int/docs/default-source/coronaviruse/situation-reports/20200123-sitrep-3-2019-ncov.pdf?sfvrsn=d6d23643_8 (accessed December 22, 2021).

WHO. (2020b). Available online at: https://www.who.int/emergencies/diseases/novel-coronavirus-2019/technical-guidance/naming-the-coronavirus-disease-(covid-2019)-and-the-virus-that-causes-it (accessed December 22, 2021).

WHO. (2020c). Available online at: https://www.who.int/emergencies/diseases/novel coronavirus-2019/question-and-answers-hub/q-adetail/q-a-coronaviruses#:~:text=symptoms (accessed December 22, 2021).

WHO. (2020d). Available online at: https://www.who.int/news-room/commentaries/detail/modes-of-transmission-of-virus-causing-covid-19-implications-for-ipc-precaution-recommendations (accessed December 22, 2021).

WHO. (2020e). Available online at: https://www.who.int/docs/defaultsource/coronaviruse/situation-reports/20200131-sitrep-11-ncov.pdf?sfvrsn=de7c0f7_4 (accessed December 22, 2021).

Yang, X. S. (2012, September). Flower pollination algorithm for global optimization. In *International Conference on Unconventional Computing and Natural Computation*. Springer: Berlin, Heidelberg, 240–249.

Chapter 9

An optimized filter design using fractional anisotropic diffusion technique for image denoising in AI environment

Subhadarshini Das, Bibhuprasad Mohanty, and Mihir Narayan Mohanty

9.1 INTRODUCTION

Biomedical image filtering techniques are essential tools used in medical image processing to enhance, denoise, segment, or otherwise manipulate images obtained from various medical imaging modalities such as MRI (Magnetic Resonance Imaging), CT (Computed Tomography), ultrasound, and microscopy. These techniques play a crucial role in improving image quality, extracting relevant information, and aiding in medical diagnosis and treatment. Some biomedical image filtering techniques are given as follows.

9.1.1 Spatial domain filtering

Mean Filtering: This technique replaces each pixel value with the average value of its neighborhood, effectively reducing noise. Whereas in Median Filtering, it replaces each pixel value with the median value of its neighborhood, robust to outliers and preserves edges. But Gaussian Filtering applies a Gaussian kernel to smooth the image, effectively reducing noise while preserving edges.

9.1.2 Frequency domain filtering

In this case, Fast Fourier Transform (FFT) can be used that converts the spatial image to frequency domain, where filtering operations can be applied more efficiently. Based on three different types of frequencies, high-pass filtering enhances high-frequency components, useful for edge detection; low-pass filtering suppresses high-frequency noise, useful for denoising; and band-pass filtering filters out frequencies outside a specified range, useful for isolating specific features.

 DOI: 10.1201/9781003589716-9

9.1.3 Anisotropic filtering

Considers in two different methods as:

- Perona-Malik Diffusion: An iterative technique that smoothers homogeneous regions while preserving edges by adaptively diffusing the image.
- Coherence-Enhancing Diffusion: Enhances structures with coherent orientation while suppressing noise and unrelated structures.

9.1.4 Non-linear filtering

In this approach, Bilateral Filtering considers both spatial closeness and intensity similarity to preserve edges while smoothing regions and Non-Local Means Filtering average pixel values based on similarity across the entire image rather than just the local neighbourhood, effective for removing noise while preserving texture and fine structures.

9.1.5 Edge enhancement

Different operators such as Sobel, Prewitt, and Roberts detect edges by calculating gradients in different directions. Whereas, Canny Edge Detector utilizes multistage algorithms to detect a wide range of edges while suppressing noise.

9.1.6 Wavelet transform

- *Discrete Wavelet Transform (DWT)*: It decomposes the image into different frequency bands, allowing for multiresolution analysis and filtering.
- *Wavelet Denoising*: Utilizes the wavelet transform to remove noise from biomedical images while preserving important features.

9.1.7 Morphological filtering

- *Erosion and Dilation*: Basic operations for removing small objects or filling gaps in structures.
- *Opening and Closing*: Combining erosion and dilation operations to remove noise and smooth object boundaries.

These techniques can be combined or customized based on the specific characteristics of the biomedical images and the requirements of the analysis or diagnosis tasks. Additionally, adaptive filtering techniques are also being increasingly utilized for biomedical image filtering tasks, offering enhanced performance and automation.

9.2 ADAPTIVE FILTERING TECHNIQUES

Adaptive filtering techniques for denoising biomedical images involve methods that adjust their filtering parameters based on the local characteristics of the image. These methods are particularly effective in scenarios where noise levels vary across different regions of the image or when noise characteristics are not constant. Here are some adaptive filtering techniques commonly used for denoising biomedical images.

9.2.1 Adaptive mean filtering

In Local Adaptive Filtering, the filter size and shape are adapted based on the local statistics of the image. This helps in preserving image details while effectively reducing noise. Similarly, in Weighted Mean Filtering, it assigns different weights to pixels within the filter window based on their similarity to the central pixel. This allows the filter to adapt to the local image structure and noise distribution.

9.2.2 Adaptive median filtering

Weighted Median Filtering is one of the techniques that assigns weights to the pixel values within the filter window based on their distance from the central pixel. This helps in preserving edges and fine details while effectively removing noise. But in Rank-Ordered Mean Filter method, it computes the mean of the pixel values within the filter window after sorting them. The weights, based on their ranks, assigned to the sorted pixel, with higher weights given to pixel values closer to the median.

9.2.3 Adaptive wiener filtering

Local Variance Estimation method estimates the local variance of the noise within the filter window and adapts the Wiener filter parameters accordingly. This helps in achieving better noise suppression while preserving image details. Spatially Variant Wiener Filtering adjusts the Wiener filter parameters based on the spatial characteristics of the image. This allows the filter to effectively handle regions with different noise levels.

9.2.4 Adaptive non-local means filtering

In Local Patch Similarity Estimation, it estimates the similarity between local image patches within the filter window and adapts the filtering parameters based on this similarity measure. This helps in preserving image textures and structures while reducing noise. But in Patch Grouping and Weighting method, it groups similar patches from different regions of the image and assigns weights to them based on their similarity. This allows the filter to adaptively combine information from different regions for better denoising.

9.2.5 Adaptive wavelet thresholding

Local Wavelet Coefficient Estimation and Adaptive Threshold Selection are two methods used. Local Wavelet Coefficient Estimation method estimates the wavelet coefficients within the filter window and adapts the thresholding parameters based on local image statistics. The Adaptive Threshold Selection selects the thresholding parameters adaptively based on the local image characteristics, such as variance or edge strength. This ensures effective noise reduction while preserving important image details.

These adaptive filtering techniques offer effective solutions for denoising biomedical images by adapting their filtering parameters based on the local characteristics of the image. By doing so, they can achieve better noise suppression while preserving important image features and structures, thus improving the quality and reliability of biomedical image analysis and diagnosis. However, the Improved Adaptive Complex Diffusion Despeckling Filter represents a sophisticated approach to denoising ultrasound images by leveraging adaptive parameter estimation, complex diffusion processes, and enhanced edge preservation mechanisms. It addresses the specific challenges posed by speckle noise in ultrasound imaging and contributes to improving the quality and reliability of medical image analysis and diagnosis.

9.3 FILTER FOR SPECKLE REDUCTION

Biomedical imaging plays a critical role in medical diagnosis and research, providing valuable insights into biological structures and processes. However, biomedical images are often corrupted by speckle noise, a granular artifact that arises because the imaging process is coherent. Speckle noise can degrade image quality, obscure fine details, and impede accurate interpretation. Hence, the development of effective speckle reduction filters is essential for improving the reliability and utility of biomedical imaging.

A despeckling filter, also known as a speckle reduction filter, is a type of image processing technique used to reduce or remove speckle noise from images. Speckle noise commonly occurs in images produced by ultrasound imaging systems, synthetic aperture radar (SAR), and certain types of microscopy. Speckle noise manifests as granular patterns that can obscure fine details and degrade image quality. Despeckling filters aim to mitigate these effects while preserving important image features. Median Filter is one of the simplest and most widely used despeckling filters. It replaces each pixel value with the median value within a local neighborhood. It is effective for removing speckle noise while preserving edges and fine details. Lee Filter is a statistical filter that achieves speckle reduction by averaging pixel values within a local window, weighted based on the signal-to-noise ratio. Anisotropic Diffusion is based on partial differential equations; this filter diffuses image intensities while preserving edges. It utilizes diffusion coefficients that are adapted based on image gradients and effective for reducing speckle noise while preserving image structures.

Despeckling filters play a crucial role in improving the interpretability and diagnostic accuracy of images affected by speckle noise. The choice of filter depends on factors such as the characteristics of the noise, the desired level of noise reduction, and the specific requirements of the imaging application.

Designing a despeckling filter for biomedical image denoising requires careful consideration of the characteristics of the noise, the specific features of biomedical images, and the desired level of noise reduction while preserving important image details. An approach for designing such a filter is given.

9.3.1 Understanding biomedical image characteristics

Biomedical images often exhibit specific features such as textures, edges, and fine structures that need to be preserved during denoising. The imaging modality (e.g., MRI, ultrasound, microscopy) and the typical noise patterns associated with it are to be considered.

9.3.2 Analysis of speckle noise

Speckle noise often exhibits multiplicative properties, where noise variance is proportional to the signal intensity. It is necessary to consider whether the speckle noise is homogeneous or exhibits variations across different regions of the image.

Then choose an appropriate filtering technique based on the analysis of speckle noise characteristics and the specific requirements of biomedical image denoising. If adaptive filtering techniques are used, develop methods for estimating filter parameters adaptively based on local image characteristics. Noise variance, signal-to-noise ratio, or texture features adaptively adjust filter parameters. By following these steps, you can systematically design a despeckling filter tailored to the specific requirements of biomedical image denoising, leading to improved image quality and reliability for diagnostic and analytical purposes.

9.4 RELATED LITERATURE

Filtering of signal and image in the field refers to preprocessing work. It needs to remove the noise and prepare the data perfectly for better and fast processing. Further, different filters have to be designed based on the type of input that may be signal or image. The work focuses on image filtering and smoothening to perform better at the time of processing. Some of these works are cited in this section.

Biomedical data acquires noise at the time of signal and image acquisition due to external devices, electrical disturbances, and also physiological disturbances.

Different signals have been proposed by authors for individual and mixed signal filtering (M. D. Mohanty et al., 2022). Similarly, the use of filters was proposed by researchers for speech and acoustic signals. These filters were based on adaptive algorithms. Further, filters were used based on adaptive algorithms and utilized for biomedical signals (Barik et al., 2018; Ram & Mohanty, 2018).

Adaptive filter design occupied a significant position in the application of biomedical data denoising. The works included brain image using intelligent filter design (M. N. Mohanty & Panda, 2017; Panda & Mohanty, 2016, 2018), X-ray images (Das et al., 2022), CT and MRI images (Dehuri et al., 2015; Jyoti et al., 2014, 2015; Kar & Mohanty, 2020), and found suitable for further processing. The adaptability depends on adaptive algorithms, neural networks, and machine learning approaches.

Systems for biomedical image denoising are essential for precise clinical diagnosis. This work aims to propose an efficient and straightforward multistep iterative image denoising method based on Wiener filtering (WF), in which the input for each stage is the previously processed image. When an adaptively reached state, as indicated by image energy, is met, the denoising process comes to an end.

Real clinical images are used in Lahmiri (2017) to test the suggested iterative method, and the performance is gauged using the widely recognized peak-signal-to-noise-ratio (PSNR) metric. According to experimental results, the suggested iterative system performs better than traditional image denoising algorithms, such as the baseline Wiener filter, wavelet packet (WP), fourth-order partial differential equation (FOPDE), non-local Euclidean means (NLEM), first-order local statistics (FOLS), and so on. The experimental findings show that the suggested method may efficiently and automatically eliminate noise while maintaining the integrity of edges and texture features.

In the study by Behera and Mohanty (2022), the authors considered the EEG signal contaminated with ocular artifacts. The clean EEG as well as artifactual EEG were taken from the openly available Mendeley database for performance verification of the proposed algorithm. The Deep Wavelet Vector Functional Link Network (WVFLN) model was used because of the non-linearity and non-stationarity nature of the signals. An exponentially weighted Recursive Least Square (RLS) algorithm was used for weight updation of the output layer.

Enhancing the quality of CT images requires the use of edge-preserving denoising techniques. Noise reduction and the retention of real, medically significant content are mutually exclusive, nevertheless. It's difficult to reduce noise without sacrificing the image's critical details, such edges, corners, and other sharp structures.

A summary of some important work in the field of CT image denoising can be found in Diwakar and Kumar (2018). Researchers frequently struggle to interpret the noise in CT scans and to choose the best denoising technique for their needs. Thus, a succinct overview of CT imaging, the features of image noise, and widely used techniques for denoising CT images were given.

Assuming maximum self-similarity, Bansal et al. (2014) proposed PCA_NLM approach for similar-neighbor computation after PCA projection. The accuracy improves and computational cost decreases by computing PCA projection to a lower dimensional subspace. The six different parameters exhibit enhanced performance when compared with standard and benchmark methods.

Mainly two approaches are used for de-noising of biomedical images: the first is filter based and the second is wavelet based. In a reported survey (Avhad & Patil, 2016), filter-based and wavelet-based technique provides satisfactory results.

For optimal image filtering, an integrated general technique based on Hebbian learning has been proposed by Mondal et al. (2006) to develop a finite impulse response filter. By employing Hebbian learning to update the filter coefficient, this approach takes use of the interpixel connection. In order to get effective learning from the neighboring pixels, the method was made iterative. This technique preserves both high-frequency and low-frequency features to get the best possible smoothing of the noisy image. The evaluation findings demonstrate the robustness of the proposed finite impulse response filter against a variety of noise distributions, including speckle, salt-and-pepper, and Gaussian noise.

A reliable wavelet domain technique for noise reduction in medical image processing was presented by Pizurica et al. (2003). A single parameter can be used to balance the degree of noise reduction against the preservation of significant features that are depending on the expertise, while also accommodating different types of image noise. In order to perform a preliminary coefficient classification, the method takes advantage of generally reliable knowledge about the correlation of relevant image features across the resolution scales. The statistical distributions of the coefficients that, on the one hand, reflect valuable image features and, on the other, primarily noise are empirically estimated using this early classification of the coefficients. A wavelet domain indicator of the local spatial activity is used to adjust the image to its spatial context. The proposed approach has a low level of complexity in terms of execution time and implementation. The outcomes show how beneficial it is for magnetic resonance imaging and medical ultrasonography noise reduction. The suggested method performs significantly better in these applications than single-resolution spatially adaptive algorithms, both in terms of the visual quality of the images and quantitative performance measures.

Ali (2018) reported a variant of median filter implementation as applied to MRI images after being added with noises gradually, such as Gaussian, salt-and-pepper, etc.

Discrete cosine transform (DCT)-based filter has been proposed by P. K. Gupta and Kanhirodan (2006) which sorts the coefficient in multiresolution-like structure. Such re-arrangement provides optimal smoothing, better energy clustering, and robust to noisy environment.

A robust polynomial filter was used by Bhateja et al. (2013) to create a non-linear framework that enhanced mammographic abnormalities resulting from biomedical instrumentation, such as X-ray instrumentation. The method described in this work achieved enhancement of mammographic masses and calcifications regardless of the kind of background tissues by using a linear combination of Type-0 and Type-II polynomial filters as a generalized filtering solution. A Type-0 filter reduces the negative effects of background noise while enhancing contrast. Conversely, Type-II filter preserves finer details by performing edge augmentation. The region of interest's degree of contrast improvement is measured using the Contrast Improvement Index, a performance metric. Furthermore, signal-to-noise ratio estimate is done using PSNR and ASNR. Furthermore, signal-to-noise ratio (PSNR and ASNR) estimation is performed to take into consideration the over-enhancements of the processed mammograms and the suppression of background noise levels. These metrics provide a way to choose the filter settings as best as possible. They also act as a quantification platform for evaluating how well the suggested filter performs in comparison to other non-linear enhancement methods that may be applied to a range of biomedical image sensors.

By utilizing image-dependent threshold parameter, Gupta et al. (2021) proposed a new diffusion coefficient. The algorithm indicates fast convergence than those found in literature.

An BM3D noise processing approach for the diffusion equation was presented by Zhang and Sun (2020) in order to reduce picture noise while maintaining image details, particularly at the edges. This approach first preprocesses images using anisotropic diffusion (AD) filtering and then searches for similar blocks using the edge direction rather than the horizontal direction. The primary goal of improving the AD model is to obtain more precise edges and intricate processing effects. First, a 5×5 edge enhancement operator model is built in eight directions using the enhanced AD direction, and the appropriate gradient information is obtained. This operator achieves good continuity and distinct outlines by optimizing the processed picture edges. The diffusion function's coefficient is then generated using a hyperbolic tangent function, and a new technique of calculating it is then shown. It is mathematically demonstrated that the diffusion function of the suggested approach, which is based on the relationship between the diffusion function and image gradient, converges more quickly than the diffusion function of the Perona and Malik model. According to the experimental results, the new model may significantly improve both the objective numerical indicators and the subjective visual effects while successfully retaining the image edges and texture features and preventing edge ringing.

For an accurate segmentation of color images with substantial noise, Maeda et al. (2014) developed a modified anisotropic diffusion algorithm that uses dynamic morphological filtering as a new exact edge-preserving smoothing technique. To enhance edge-preserving smoothing, they added a dynamic selection of several structural elements for every pixel in the morphological filter prior to anisotropic diffusion. Consequently, prior to executing a segmentation technique, noise and superfluous information in images are efficiently smoothed while maintaining small-image structures. A number of simulated examples are provided to show how effective the proposed method is.

The initial steps toward vision processing must deal with random intensity variations, illumination variations, or low contrast which have the potential to contaminate the image. To solve the issue, the current approach introduces a weighted guided image filter by combining edge-aware weighting into a guided image filter.

Krishna and Valluru (2016) presented the derivation of an imaging-specific diffusion approach called noise-reducing anisotropic diffusion. It is possible to eliminate noise from digital photos using anisotropic diffusion without causing edge blurring. The anisotropic diffusion equations reduce to the heat equation, which is the same as Gaussian blurring, with a constant diffusion coefficient. The experimental findings demonstrate that the suggested approach outperforms the state-of-the-art in terms of criteria.

Considering diffusivity as a tensor valued function, Weeratunga and Kamath (2003) extended the existing anisotropic diffusion for suitable local edge orientation. This allows smoothing along the edges, but not perpendicular to it. They investigated and proved that their approach is relatively much better than isotropic non-linear diffusion-based techniques.

The efficiency with which various edge stopping functions preserve edges in anisotropic diffusion was examined by Kamalaveni et al. (2015). They discovered that sharp edges and minute details are well preserved using an edge stopping function that effectively halts diffusion from low image gradient onward. In the case of level set techniques, this characteristic of an edge halting function led to a lower evolution. However, since fine features and sharp edges are blurred by diffusion, an edge stopping function that halts diffusion from a high picture gradient onward would not preserve them. They discovered that, when applied to the edge stopping function, low gradient threshold parameter values better maintain fine features and sharp edges than high gradient threshold parameter values.

Ovireddy and Muthusamy (2014) proposed and applied a speckle suppressing anisotropic diffusion (SSAD) filter to B-mode ultrasound image and also tested for its efficacy over polycystic ovary images available from clinics. Both the functional- and numerical parameter-based evaluation provides better results.

In the study by Tsiotsios and Petrou (2013), the optimal option is chosen after a variety of different possibilities are evaluated and assessed at each step of the algorithm. An automatic stopping criterion that considers not only the attained smoothing level but also the quality of the maintained edges has been proposed. With the aid of real and synthetic images, the suggested plan was assessed and contrasted with other cutting-edge systems according to objective standards.

An approach that is modified from the anisotropic diffusion filter (ADF) is shown in the study by Kumar et al. (2020). It uses the ADF regularization as a transitional step to smooth the MRI image. It is a more thorough and refined form of 'Perona-Malik' Diffusion that, when compared to alternative techniques, yields a higher image quality evaluation. Peak signal-to-noise ratio (PSNR), mean square error (MSE), and structural similarity index (SSIM) are among the picture quality assessment criteria that are calculated in relation to the complete reference image. Some metrics without a reference, such as the Blind/Reference-less Image Spatial Quality Evaluator (BRISQUE) and the Perception-based Image Quality Evaluator (PIQE), are also established. The estimated values for the whole reference image for the Kaggle data set are 39.13dB, 7.21, and 0.99, in that order. The average results regarding for no reference metric are given as 48.84 and 37.13, respectively.

In the study by Abuya et al. (2023) additive Gaussian blur noise (AGBN) is eliminated in CT scan images. They use an ensemble approach based on deep learning denoising with anisotropic Gaussian filter (AGF) and wavelet transform. Due to non-uniform noise distribution in CT scan images, AGF provides better directional information and edge orientation.

Black et al. (1998) established that given a noisy image, anisotropic diffusion can estimate a piece-wise smooth image robustly. They proposed the edge-stopping function based on Turkey's weight for preservation of boundaries. It also helps in improving the automatic stopping of the diffusion.

Teaching-Learning-Based Optimization (TLBO) methods to achieve the targeted performance analysis of the denoising mechanism have been proposed by Kushwaha and Singh (2017). But it incurred loss of data and overfiltering in spite of its effectiveness.

Rajkumar (2014) proposed the inclusion of Vector Median Filtering (VMF) to improve their performance when dealing with multimodal noises. They included multiscale edge detector into the formulation of the Perona-Malik (PM) anisotropic diffusion scheme. This facilitated the scheme of various noises removal such as Gaussian, impulse and photon noise.

To restore and denoise Satellite Remote Sensing Images (SRSIs), Gatcha et al. (2022) proposed the motion by a mean curvature approach to detect the noise direction for each degraded pixel. The algorithm preserves the gradient in the Gaussian kernel and saves the contrast of the image.

Chao et al. (2010) proposed a method which can simultaneously preserve edges and fine details while filtering out noise in the diffusion process. The proposed diffusion model incorporates both local gradient and gray-level variance to preserve edges and fine details while effectively removing noise. As compared to all other existing techniques, this method has the best restoration results.

Based on GPU accelerate technique and subpixel anisotropic diffusion, Guo and Wei (2015) improved an image removal algorithm. GPU implementation plays an important role in alleviating time costs. Better performance and less time consumption is achieved through the algorithm.

Yuan (2015) modified the Perona-Malik model by introducing gradient information, intensity and space position in the new non local-mean (NLM) theory. In the proposed model, the intensity, space position, and gradient information are introduced into NLM model, which weakens the stair casing effect but preserves edge in a processed image. The new PM model enhances the denoising capability and decreases fine characteristics to be over-smoothed. Comparative experiments show that the denoising capability of the proposed model is superior to the other three existing models.

Harrabi and Braiek (2012) present a comparative analysis of picture denoising methods. The authors extended the technique into regular anisotropic diffusion after analyzing the inefficiency of isotropic and anisotropic diffusion. Total variation-based diffusion is used along probable edges, while isotropic diffusion is used at low gradient sites. To demonstrate the process, these denoising algorithms have been applied to satellite and textured images. The classification accuracy from various denoising approaches is verified, and the PSNR for the available test data is assessed. The experimental findings show that regular anisotropic diffusion is optimal for picture denoising.

The authors talked about various non-linear methods for image denoising or noise removal. The algorithm design, goal, efficacy, and efficiency of these methods vary. The article provides a brief description of several denoising techniques along with a discussion of their benefits and drawbacks. Statistical approaches are used to analyze and compare the performances of these techniques based on various implementations. Lastly, the study concludes what kinds of noise each denoising technique can efficiently manage and when it works best (Abdalla et al., 2018).

In the study by Liu and Liu (2012), anisotropic diffusion is taken into account for image denoising. Diffusion tensor construction is approached in a novel way. Our method yields tensors that are adaptively determined by local image structure, as they rely on four directional derivatives of an image's intensity. It is demonstrated that the suggested diffusion filter is anisotropic at the borders of a region and isotropic in the center. We can effectively eliminate noise in an image, especially noise at the edges, thanks to this tensor characteristic. Both synthetic and real images are used in a number of numerical investigations.

In the study by Hadj Fredj and Malek (2021), the authors demonstrated the high-level synthesis implementation of an anisotropic diffusion algorithm for medical purposes using a SOC-FPGA. The oriented speckle reduction anisotropic diffusion denoising filter that the authors have selected offers strong performance but necessitates a large amount of work on the integrated CPU due to its iterative nature. By altering the original method, automating it by managing the diffusion process at each iteration, and speeding up processing activities by offering a hardware/software description, they were able to maximize performance. Various medical photos were used for the examination. Segmentation is used to illustrate the effectiveness and applicability of the suggested filter. The FPGA XC7Z020CLG484-1 is used to validate the design at a frequency of 255 MHz and a PSNR of roughly 30 dB.

One of the best techniques for processing images is anisotropic diffusion. It can be used to remove an image's minor textures while keeping its important borders intact.

To increase the filter's overall performance, a new anisotropic diffusion filter based on a fractional calculus kernel—as opposed to an integer kernel—is developed. Utilizing the advantages of parallelism in FPGAs, the Genesys-2 FPGA kit is used to create integer and fractional anisotropic filters. The feasible PSNR value vs the number of repetitions is used to evaluate integer and fractional anisotropic filters. In comparison to an integer anisotropic filter, the suggested fractional anisotropic filter reduces power and area while achieving a higher PSNR value with fewer iterations. The suggested filter can be applied to cartooning, image denoising, segmentation, edge detection, and smoothing. Additionally, because the suggested filter needs fewer iterations than integer anisotropic filters, it will conserve power in situations with power limits, like portable devices. The suggested filter lowers the necessary power usage by 58.2%.

The two main categories of images are anisotropic and isotropic. Anisotropic images are composed of structures that resemble flows, whereas isotropic images mostly feature spherical objects. Noise is introduced during the acquisition process, regardless of the type. Diffusion is a common method for image smoothing. Diffusion can be either isotropic or anisotropic, depending on the category. Diffusion is essentially an iterated process that begins with a low-quality image and ends with a mean-value image that is entirely blurred and devoid of any discernible structure. Although the process begins with under-smoothing, which is a desirable task of filling in gaps and removing noise, it soon moves into an over-smoothing phase where it begins to destroy the crucial structure. Determining the border between the under-smoothing and over-smoothing zones is one pertinent issue. One such metric that could be useful in offering crucial hints to characterize that border is the spatial entropy change, which offers a logical stopping rule for both isotropic and anisotropic diffusion. The role of entropy-change in determining an appropriate stopping point where the majority of the noise is reduced and blurring has just begun is confirmed by numerical experiments using actual fingerprint data. The blurring phenomenon, which is a growing function of the diffusion process, is directly related to the suggested criterion.

The suggested method is assessed using both synthetic and actual photos in AbdAlRahman et al. (2023) and contrasted with other cutting-edge techniques using a qualitative metric. The suggested stopping rule is used to establish a plausible stopping rule based on the analysis of the diffusions of a few difficult low-quality images from FVC2004. In image processing, it can be difficult to smooth images while maintaining edges when there are outliers. In digital image smoothing with edge preservation, anisotropic diffusion smoothing is a well-known paradigm. Diffusion smoothing filters that are anisotropic are not resistant to impulsive noise. They require far more arithmetic operations to be robust in the presence of outliers as a result of using two or more stages; as a result, they are not power efficient.

In the study by Khan et al. (2016), an approach for anisotropic diffusion smoothing of images that is both fundamentally resilient and power efficient has been developed, along with a low arithmetic complexity image smoothing model. The technique performs better than the fundamental robust smoothing algorithms in terms of visual quality and conventional performance measures. Nair et al. (2019) suggested the use of such multiscale anisotropic diffusion to denoise noisy photos. Digital images typically contain objects of various sizes, and one of the primary goals in early computer vision problems is to denoise them without erasing their edges. They made use of the information found in the numerous scale-based images created by anisotropic PDE, in contrast to earlier methods that ignore them. The suggested system outperforms previous schemes in terms of noise reduction and detail preservation by skillfully integrating the interscale details.

Anisotropic diffusion (PM) and total variation (TV) models were used to create a novel image edge identification method. The developers of the suggested method first employed the gradient magnitude to remove noise from the image before using the adaptive weight function to identify the image's edges. The adaptive weight function is highly capable of changing and adapting to the information about the areas (flats or edges). More precisely, areas with weak and irregular edges are treated with anisotropic diffusion filters, whereas areas with double and false edges are treated with TV filters. The majority of the false edges can be eliminated by applying TV filters to the double-edged areas, resulting in significantly sharper edges. On the other hand, robust and continuous edges can be obtained by applying an anisotropic diffusion filter to the areas with discontinuous edges. As a result, in addition to great localization accuracy, there were fewer erroneous edges. Results from experiments show that the new method is superior in terms of eliminating erroneous edges and enhancing edge localization accuracy. Pratt's figure of merit (FOM) and the peak signal-to-noise ratio (PSNR) were employed as objective quantitative performance metrics (Prasath, 2017).

Three key elements in radiographic imaging—contrast, sharpness, and noise—determine the quality of the image. It can be challenging to eliminate noise while maintaining and enhancing image outlines, especially for low-contrast images like radiography.

In the study by Abdullah Yahya et al. (2019), a novel anisotropic diffusion technique for improving radiography images has been put forth. The suggested technique, which may simultaneously carry out the smoothing and sharpening operations, was based on the integration of geometric parameters obtained from the local pixel intensity distribution in a non-linear diffusion formulation. The ability to integrate edge preservation, sharpening, and noise reduction into a single process was the primary innovation of the suggested anisotropic diffusion model. The effectiveness of the suggested approach in contrast to alternative anisotropic diffusion techniques is demonstrated by experimental findings utilizing both synthetic and actual welding radiography pictures.

9.5 METHODOLOGY

Anisotropic diffusion (AD) is a powerful technique for image denoising that preserves edges and details while reducing noise. The basic concept of AD is to diffuse the image intensity over time, with diffusion rates that depend on the image gradient. Here's mathematical overview of how AD works for image denoising. Biomedical images often possess both magnitude and phase information, making complex representation suitable. A complex-valued image $I(z)$ is represented as: $I(z) = M(z) \cdot ej\Theta(z)$, where $M(z)$ is the magnitude and $\Theta(z)$ is the phase at each pixel location z. The diffusion coefficient c is often adaptively estimated based on local image characteristics. This is done using statistical measures such as local variance, image gradients, or texture features. Adaptive estimation ensures that the filter parameters are optimized for different regions of the image. The diffusion equation is typically solved numerically using finite-difference methods or other numerical techniques. Discretization of the equation yields a system of equations that can be solved iteratively to obtain the denoised complex image. The diffusion coefficient is a non-negative function of the image gradient magnitude that decreases monotonically. As a result, the blurring impact will be minimal near the edge regions where the gradient magnitude is big since the diffusion coefficient is small. In image processing and computer vision, anisotropic diffusion is an optimal method for lowering image noise while maintaining crucial image features like edges and lines. It is based on a partial differential equation (PDE) and is also known as Perona–Malik diffusion. PM flux for regions where the gradient magnitude is higher than smoothing threshold may lead to undesirable blurring effect and edge displacement. The diffusion coefficient controls the rate of diffusion and may be adaptive to preserve edges and details.

9.5.1 Diffusion Equation

The basic diffusion equation for AD is given by:

$$\frac{\partial I}{\partial t} = \nabla\left(C\|\nabla I\|\right)\nabla I)$$

where

I is the image intensity,
t is time,
∇ is the gradient operator,
$\|\nabla I\|$ is the magnitude of the image gradient, and
$C(\|\nabla I\|)$ is the diffusion coefficient that controls the rate of diffusion based on the gradient magnitude.

9.5.2 Diffusion Coefficient

The diffusion coefficient $C(\|\nabla I\|)$ is a function of the gradient magnitude and determines how quickly diffusion occurs in different parts of the image. It is typically defined as:

$$C(\|\nabla I\|) = e^{-\left(\frac{\|\nabla I\|}{k2}\right)^2}$$

where k is a parameter that controls the strength of diffusion. A larger k value results in less diffusion and more edge preservation.

In FA, the mathematical representation of a noisy image undergoing denoising can be described by a partial differential equation (PDE) that models the diffusion process. The PDE typical takes the form of a fractional diffusion equation.

Let's denote the noisy image as $I_{\text{noisy}}(\chi,\gamma)$ and the denoised image as I_{noisy} (χ,γ). Then, the FAD process can be represented by the following partial differential equation:

$$\frac{\partial I_{\text{denoised}}}{\partial t} = \nabla.(C(x,y)\left|\nabla I_{\text{denoised}}{}^{m-1}\right|\nabla I_{\text{denoised}}$$

where

$\frac{\partial I_{\text{denoised}}}{\partial t}$ represents the rate of change of the denoised image with respect to time.
∇ represents the gradient operator.
$\left|\nabla I_{denoised}\right|$ represents the magnitude of the gradient of the denoised image.

m is the fractional order parameter.
$C(\chi,\gamma)$ is a spatially varying diffusion coefficient which can be adaptively determined based on local image characteristics.

This equation describes now the denoised image evolves over time by diffusing the image intensities in a manner that is biased by the local image structure and noise characteristics. The fractional order parameter allows for controlling the degree of smoothing and edge preservation in the denoising process.

$$\frac{\partial^{\alpha} I}{\partial t_{\alpha}} = \nabla.\left(\mathrm{C}(\|\nabla I\|)\nabla I\right)$$

where

I is the image intensity,
t is time,
∇ Is the gradient operator,
$\|\nabla I\|$ is the magnitude of the image gradienof the image gradient,
$\frac{\partial^{\alpha} I}{\partial t^{\alpha}}$ Is the fractional derivative of order α.
$\mathrm{C}(\|\nabla I\|)$ is the diffusion coefficient, and
$0 < \alpha < 1$ is the fractional order and the optimized value is 0.89.

9.5.3 Fractional Derivative

The fractional derivative in the fractional diffusion equation is defined in terms of the Riemann Liouville fractional derivative:

$$\frac{\partial^{\alpha} I}{\partial t_{\alpha}} = \frac{1}{\Gamma(1-\alpha)}\frac{\partial}{\partial t}\int_{0} t(t-T)^{-\alpha}\frac{\partial I}{\partial T}\,\mathrm{d}\tau$$

where Γ is the gama function.

9.5.4 Diffusion Coefficient:

The diffusion coefficient $\mathrm{C}(\|\nabla I\|)$ in FAD controls the rate of diffusion based on the image gradient, similar to AD. It is typically defined as:

$$\mathrm{C}(\|\nabla I\|) = e^{-\left(\frac{\|\nabla I\|}{k2}\right)^{2}}$$

where κ is a parameter that controls the strength of diffusion.

9.6 RESULTS

In the proposed work, the biomedical image is given as an input as in Figure 9.1 and subjected to normalization. The normalized image is segmented (Figure 9.2) and filtering is applied. The resultant filtered image is subjected to FAD to find out the smooth image. The details are as follows.

Fractional Adaptive Complex Diffusion Despeckling Filter offers significant advantages for denoising biomedical images, including preservation of image features, edge preservation, adaptive parameter estimation, versatility across imaging modalities, and enhanced image interpretability.

Anisotropic diffusion (AD) and fractional anisotropic diffusion (FAD) are powerful techniques for image denoising, particularly in the field of biomedical imaging. These methods preserve essential image features while effectively reducing noise, making them invaluable for tasks such as MRI, CT, and ultrasound image processing.

AD utilizes the image gradient to control the diffusion process, ensuring that edges and fine details are maintained while noise is smoothed out. This is achieved through a diffusion coefficient that adapts based on local image characteristics, providing an optimal balance between noise reduction and edge preservation. The mathematical formulation of AD, with its gradient-based diffusion coefficient, allows for effective noise reduction while preserving crucial structural information in the images.

FAD extends the capabilities of traditional AD by incorporating fractional calculus into the diffusion process. This extension allows for enhanced adaptability to complex noise patterns and finer control over the degree of smoothing and edge preservation. The fractional order parameter in FAD provides an additional degree of freedom, enabling more precise tuning of the denoising process to suit specific image characteristics and noise profiles. The result is improved image quality, particularly in challenging scenarios where traditional denoising methods may fall short.

The proposed fractional diffusion filter design approach, which incorporates both AD and FAD principles, demonstrates significant potential for improving image quality in biomedical imaging. By focusing on specific regions of the image and adapting the diffusion coefficients based on local

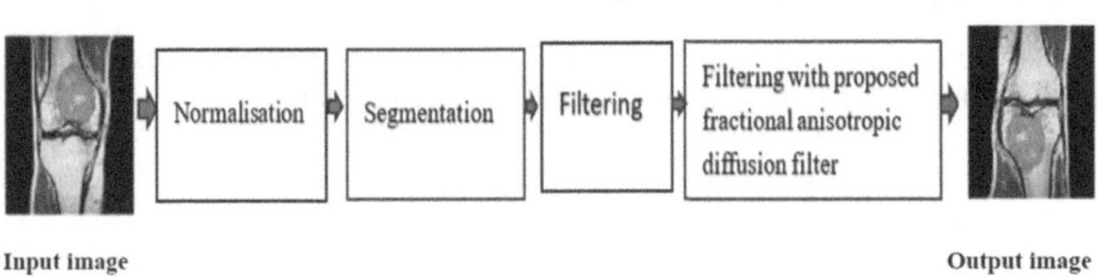

Figure 9.1 The proposed methodology for the FAD.

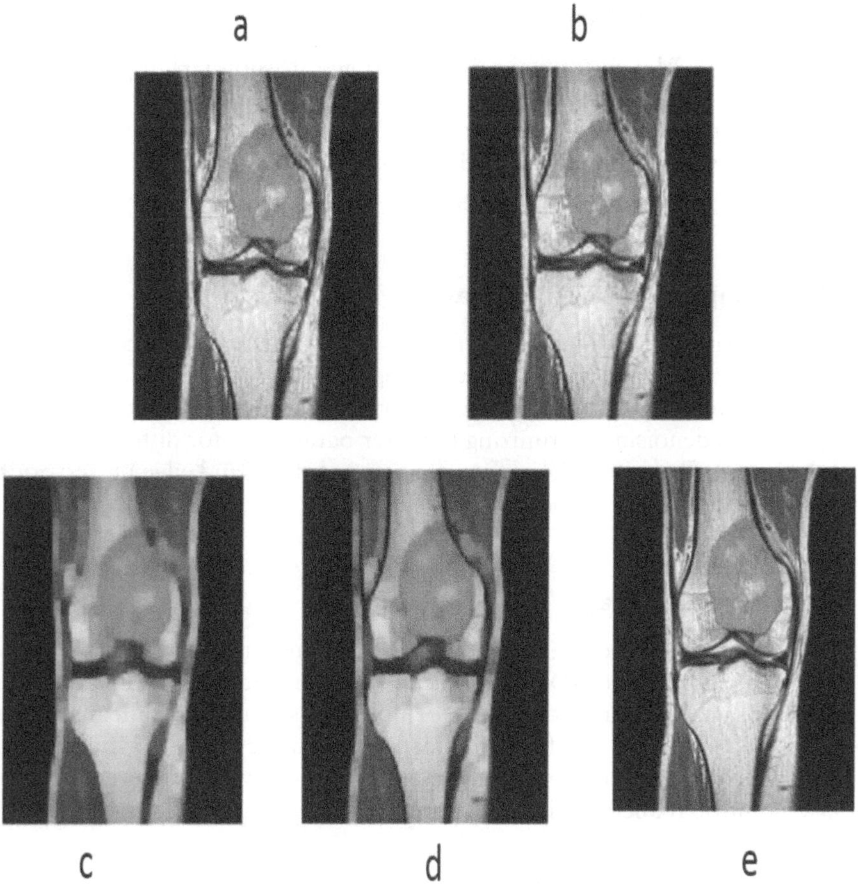

Figure 9.2 (a) Input image, (b) segmented image, (c) adaptive filtered image, (d) filtered diffused image, (e) smooth or FAD image.

image characteristics, this method effectively reduces speckle noise and enhances image sharpness and edge clarity.

Overall, the combination of AD and FAD offers a robust framework for image denoising, preserving fine structures and details while effectively removing noise. This makes them particularly valuable in medical diagnostics, where clear and accurate imaging is critical for accurate analysis and diagnosis. Future research and development in this area may further refine these techniques, leading to even more advanced and efficient denoising solutions for a wide range of imaging applications. It is taken a systematic approach to image denoising by examining three distinct methods: filtered diffusion image, adaptive diffusion image, and smooth image. Each method contributes uniquely to the overall goal of enhancing image quality while preserving essential features.

9.6.1 Filtered diffusion image

Initially, we employed traditional anisotropic diffusion techniques to filter the images. This method effectively reduces noise while preserving significant edges and details. The anisotropic diffusion process adapts the diffusion rate based on the image gradient, ensuring that edges are maintained while smoothing homogeneous regions.

9.6.2 Adaptive diffusion image

Next, we introduced adaptive diffusion, where the diffusion coefficient is dynamically estimated based on local image characteristics such as variance, gradients, or texture features. This adaptation allows for a more targeted approach to denoising, optimizing the filter parameters for different regions of the image. This results in improved noise reduction and edge preservation compared to traditional methods.

9.6.3 Smooth image

Finally, we considered the smooth image obtained through the fractional anisotropic diffusion (FAD) technique. FAD incorporates fractional calculus to provide enhanced adaptability to complex noise patterns while maintaining image structures. The fractional order parameter allows for fine-tuning the degree of smoothing and edge preservation, achieving a superior balance between noise reduction and detail preservation.

9.6.3.1 Peak signal to noise ratio

The ratio of the maximum possible signal to mean squared error is called PSNR. In this case, the PSNR between the input image and the unsharp masking image is computed in decibels (dB) and is expressed mathematically in

$$PSNR = 10\log 10\lfloor (\max^2)/MSE \rfloor$$

where the max value can be 255.

$$\text{Where MSE} = \frac{1}{(mXn)} \sum_{x=0}^{m-1} \sum_{y=0}^{n-1} \left[I0(x,y) - Ir(x,y) \right]^2$$

where mXn is the size of the image and $I0$ and Ir are the input and restored image, respectively.

9.6.3.2 Structural Similarity Index (SSIM)

$$\mathrm{SSIM}(\mathrm{I0},\mathrm{Ir}) = (2\mu x\mu y + \alpha)(2cov(I0.Ir) + \beta) / (\mu x^2 + \mu y^2 + \alpha)(\sigma x^2 + \sigma y^2 + \beta)$$

Where μx and μy are the means and σx and σy are the variances of the images $I0$ and Ir respectively, cov($I0$,Ir) is the covariance of the two images, and α, β are small values and used to stabilize the division with small denominators. SSIM is used to show the similarity of the structure of the input image to the structure of the restored image. The SSIM lies between 0 and 1. The higher value of SSIM shows a great structure similarity between input and restored image.

The segmented image (Figure 9.2b) is filtered through three different types of filters before applying the FAD. Figure 9.2(c) visually represents the effect of adaptive filtration, which gives a hazy and blurred output, thereby missing out the important information suitable for the diagnosis. The PSNR of the image was found to be 20.47 and SSIM is 0.54. Even though the filtered diffuse image has some visual improvisation over the adaptive one, still it is not capable of providing the important attributes and hence also fails for better diagnosis. The PSNR of the image was found to be 37.13 and SSIM is 0.72. But the result of FAD filtered image is superior in the sense of qualitative and quantitively. The PSNR of the image was found to be 81.79 and SSIM is 0.94.

9.7 CONCLUSION

Biomedical image filtering and sharpening are essential processes for enhancing the quality and diagnostic value of biomedical images. By effectively reducing noise, artifacts, and blurring while preserving important anatomical details and enhancing contrast, these techniques contribute to improved accuracy in medical diagnosis, treatment, and research. These techniques are used for tasks such as noise reduction, artifact removal, contrast enhancement, and feature extraction, ultimately improving the accuracy of disease diagnosis, treatment planning, and biomedical research. Continued research and development in this field will further advance the capabilities of biomedical imaging systems and improve patient outcomes.

REFERENCES

Abdalla, A. M., Osman, M. S., AlShawabkah, H., Rumman, O., & Mherat, M. (2018). A review of nonlinear image-denoising techniques. *2018 Second World Conference on Smart Trends in Systems, Security and Sustainability (WorldS4)*, 96–100.

AbdAlRahman, A., Al-Atabany, W. I., Soltan, A., & Radwan, A. G. (2023). High-performance fractional anisotropic diffusion filter for portable applications. *Journal of Real-Time Image Processing*, *20*(5), 98.

Abdullah Yahya, A., Tan, J., Su, B., Liu, K., & Hadi, A. N. (2019). Image edge detection method based on anisotropic diffusion and total variation models. *The Journal of Engineering*, *2019*(2), 455–460.

Abuya, T. K., Rimiru, R. M., & Okeyo, G. O. (2023). An image denoising technique using wavelet-anisotropic gaussian filter-based denoising convolutional neural network for CT images. *Applied Sciences*, *13*(21), 12069.

Ali, H. M. (2018). MRI medical image denoising by fundamental filters. *High-Resolution Neuroimaging-Basic Physical Principles and Clinical Applications*, *14*, 111–124.

Avhad, S., & Patil, R. A. (2016). De-noising techniques for biomedical images. *International Journal of Research in General Science*, 4, 706–710.

Bansal, M., Devi, M., Jain, N., & Kukreja, C. (2014). A proposed approach for biomedical image denoising using PCA_NLM. *International Journal of Bio-Science and Bio-Technology*, *6*(6), 13–20.

Barik, A., Mohanty, M. N., & Das, K. (2018). Convergence analysis of adaptive MSFs used for acoustic echo cancellation. *International Journal of Information and Communication Technology*, *13*(2), 196–207.

Behera, S., & Mohanty, M. N. (2022). Artifact removal using deep WVFLN for brain signal diagnosis through IoMT. *Measurement: Sensors*, *24*, 100465.

Bhateja, V., Misra, M., Urooj, S., & Lay-Ekuakille, A. (2013). A robust polynomial filtering framework for mammographic image enhancement from biomedical sensors. *IEEE Sensors Journal*, *13*(11), 4147–4156.

Black, M. J., Sapiro, G., Marimont, D. H., & Heeger, D. (1998). Robust anisotropic diffusion. *IEEE Transactions on Image Processing*, *7*(3), 421–432.

Chao, S.-M., Tsai, D.-M., Chiu, W.-Y., & Li, W.-C. (2010). Anisotropic diffusion-based detail-preserving smoothing for image restoration. *Proceedings of 2010 IEEE 17th International Conference on Image Processing*, 4145–4148.

Das, A., Mohapatra, S. K., & Mohanty, M. N. (2022). Design of deep ensemble classifier with fuzzy decision method for biomedical image classification. *Applied Soft Computing*, *115*, 108178.

Dehuri, A., Sanyena, S., Dash, R. R., & Mohanty, M. N. (2015). A comparative analysis of filtering techniques on application in image denoising. *2015 IEEE International Conference on Computer Graphics, Vision and Information Security (CGVIS)*, 141–144.

Diwakar, M., & Kumar, M. (2018). A review on CT image noise and its denoising. *Biomedical Signal Processing and Control*, *42*, 73–88.

Gatcha, M., Messelmi, F., & Saadi, S. (2022). An anisotropic diffusion adaptive filter for image denoising and restoration applied on satellite remote sensing images: a case study. *Engineering, Technology \& Applied Science Research*, *12*(6), 9715–9719.

Guo, Y., & Wei, W. (2015). GPU accelerated image noise removal approach based on subpixel anisotropic diffusion. *Turkish Journal of Science and Technology*, *10*(2), 23–27.

Gupta, B., Lamba, S. S., & others. (2021). An efficient anisotropic diffusion model for image denoising with edge preservation. *Computers \& Mathematics with Applications*, *93*, 106–119.

Gupta, P. K., & Kanhirodan, R. (2006). A DCT based filtering of biomedical images. *2006 IEEE International Conference on Industrial Technology*, 2551–2556, 10.1109/ICIT.2006.372657

Hadj Fredj, A., & Malek, J. (2021). FPGA-accelerated anisotropic diffusion filter based on SW/HW-codesign for medical images. *Journal of Real-Time Image Processing*, *18*(6), 2429–2440.

Harrabi, R., & Braiek, E. Ben. (2012). Isotropic and anisotropic filtering techniques for image denoising: A comparative study with classification. *2012 16th IEEE Mediterranean Electrotechnical Conference*, 370–374.

Jyoti, A., Mohanty, M. N., Kar, S. K., & Biswal, B. N. (2015). Optimized clustering method for CT brain image segmentation. *Proceedings of the 3rd International Conference on Frontiers of Intelligent Computing: Theory and Applications (FICTA) 2014: Volume 1*, 317–324.

Jyoti, A., Mohanty, M. N., & Kumar, M. P. (2014). Morphological based segmentation of brain image for tumor detection. *2014 International Conference on Electronics and Communication Systems (ICECS)*, 1–5.

Kamalaveni, V., Rajalakshmi, R. A., & Narayanankutty, K. A. (2015). Image denoising using variations of Perona-Malik model with different edge stopping functions. *Procedia Computer Science*, *58*, 673–682.

Kar, P., & Mohanty, M. N. (2020). An intelligent approach for noise elimination from brain image. *Advanced Computing and Intelligent Engineering: Proceedings of ICACIE 2018, Volume 1*, 391–400.

Khan, T. M., Khan, U., Kong, M. A., Kong, Y., & Kittaneh, O. (2016). Stopping criterion for linear anisotropic image diffusion: a fingerprint image enhancement case. *EURASIP Journal on Image and Video Processing*, *2016*, 1–20.

Krishna, D. H., & Valluru, P. (2016). Anisotropic diffusion based image filtering. *International Journal of Scientific Engineering and Research* 5(51),10445–10448.

Kumar, R. R., Kumar, A., & Srivastava, S. (2020). Anisotropic diffusion based unsharp masking and crispening for denoising and enhancement of MRI images. *2020 International Conference on Emerging Frontiers in Electrical and Electronic Technologies (ICEFEET)*, 1–6.

Kushwaha, S., & Singh, R. K. (2017). An efficient approach for denoising ultrasound images using anisotropic diffusion and teaching learning based optimization. *Biomedical and Pharmacology Journal*, *10*(2), 805–816.

Lahmiri, S. (2017). An iterative denoising system based on Wiener filtering with application to biomedical images. *Optics & Laser Technology*, *90*, 128–132.

Liu, F., & Liu, J. (2012). Anisotropic diffusion for image denoising based on diffusion tensors. *Journal of Visual Communication and Image Representation*, *23*(3), 516–521.

Maeda, J., Harada, T., Saga, S., & Suzuki, Y. (2014). Anisotropic dynamic-morphological-diffusion for segmentation of noisy color images. *Journal of Advanced Computational Intelligence and Intelligent Informatics*, *18*(2), 204–212.

Mohanty, M. D., Pattnayak, P., & Mohanty, M. N. (2022). Design and implementation of a mixed signal filter for noise removal in raw ECG signal. In *Cognitive Informatics and Soft Computing: Proceeding of CISC 2021* (pp. 771–781). Springer.

Mohanty, M. N., & Panda, S. (2017). Impulsive noise cancellation from cardiac signal using modified WLMS algorithm. *RICE* 10, 255–260.

Mondal, P. P., Rajan, K., & Ahmad, I. (2006). Filter for biomedical imaging and image processing. *JOSA A*, *23*(7), 1678–1686.

Nair, R. R., David, E., & Rajagopal, S. (2019). A robust anisotropic diffusion filter with low arithmetic complexity for images. *EURASIP Journal on Image and Video Processing*, *2019*, 1–14.

Ovireddy, S., & Muthusamy, E. (2014). Speckle suppressing anisotropic diffusion filter for medical ultrasound images. *Ultrasonic Imaging*, *36*(2), 112–132.

Panda, S., & Mohanty, M. N. (2016). Impulsive noise cancellation from ECG signal using adaptive filters and their comparison. *Indonesian Journal of Electrical Engineering and Computer Science*, *3*(2), 369–376.

Panda, S., & Mohanty, M. N. (2018). Use of adaptive algorithm for impulsive noise cancellation. *International Journal of Internet of Things and Cyber-Assurance*, *1*(2), 104–118.

Pizurica, A., Philips, W., Lemahieu, I., & Acheroy, M. (2003). A versatile wavelet domain noise filtration technique for medical imaging. *IEEE Transactions on Medical Imaging*, *22*(3), 323–331.

Prasath, V. B. S. (2017). Image denoising by anisotropic diffusion with inter-scale information fusion. *Pattern Recognition and Image Analysis*, *27*, 748–753.

Rajkumar, G. (2014). Denoising Algorithm Based on Anisotropic Diffusion with Vector Median Filtering. *International Journal of Engineering Research & Technology (IJERT)*, *3*(2), 44–47.

Ram, R., & Mohanty, M. N. (2018). Performance analysis of adaptive variational mode decomposition approach for speech enhancement. *International Journal of Speech Technology*, *21*(2), 369–381.

Tsiotsios, C., & Petrou, M. (2013). On the choice of the parameters for anisotropic diffusion in image processing. *Pattern Recognition*, *46*(5), 1369–1381.

Weeratunga, S. K., & Kamath, C. (2003). Comparison of PDE-based nonlinear anisotropic diffusion techniques for image denoising. *Image Processing: Algorithms and Systems II*, *5014*, 201–212.

Yuan, J. (2015). Improved anisotropic diffusion equation based on new non-local information scheme for image denoising. *IET Computer Vision*, *9*(6), 864–870.

Zhang, Y., & Sun, J. (2020). An improved BM3D algorithm based on anisotropic diffusion equation. *Mathematical Biosciences and Engineering*, *17*(5), 4970–4989.

Chapter 10

Role of enhanced visual cryptography algorithm in cybersecurity

Susmita Panda, Sadisha Sasneha, Pratyush Nayak, and Amar Jyoti Nayak

10.1 INTRODUCTION

10.1.1 Overview of cryptography in cybersecurity

The function of cryptography has expanded in today's increasingly digital world from safeguarding text-based communications to safeguarding a variety of digital assets such as pictures videos and biometric information. Since digital technologies are developing at a rapid pace and sectors like government banking healthcare and forensics are depending more and more on digital systems, the need for strong and specialized encryption techniques is greater than ever. These industries handle enormous volumes of sensitive data most of which is visual in nature (such as surveillance footage identity verification photos and medical images), necessitating special security measures to preserve privacy and confidentiality. While traditional cryptographic systems work well for many kinds of data, they frequently fail to secure visual data, because it has unique characteristics and needs that text-based systems are unable to meet. Data in text and numeric formats has long been protected using conventional encryption techniques like hashing digital signatures and symmetric and asymmetric encryption. Although symmetric encryption algorithms such as the Advanced Encryption Standard (AES) provide effective encryption and decryption procedures, they mainly depend on safe key management. By employing a pair of public and private keys to address key distribution issues asymmetric encryption techniques—such as RSA—allow for safe data transfers especially during online transactions. On the other hand, digital signatures guarantee data integrity and offer authentication enabling the recipient to confirm the messages or data's origin. These cryptography systems were not made, especially to handle the complexity of visual data, which includes pictures videos and other graphical content even though they are excellent at protecting textual and numerical data during communications and data storage. Traditional cryptographic systems frequently encounter problems with visual data including pixel expansion color fidelity loss and image distortion during decryption. When an image is encrypted using traditional methods, it is usually treated

DOI: 10.1201/9781003589716-10

as just a large binary object, failing to account for the inherent structure and characteristics of the image, such as pixels and color channels. As a result, the decrypted image may be distorted or altered in a way that makes it unsuitable for practical use, particularly in fields like medical imaging, forensic analysis, and digital media, where preserving the original quality of the image is critical. The idea of visual cryptography which entails dividing an image into several pieces (or shares) that can only be pieced back together to form the original image is also not intrinsically supported by these systems. By making sure that no single share alone contains significant information, this method adds an extra layer of security. It is especially helpful for safeguarding sensitive visual data. Here is where visual cryptography comes in handy providing a special method created, especially to meet the security and integrity requirements of visual data. By dividing the original image into shares, Visual Cryptography Schemes (VCS) make safe image sharing possible. According to Naor et al., these shares may be non-meaningful and appear as random noise or they may have discernible patterns. On their own these shares seem like random noise or meaningless pictures, but they can be superimposed to recreate the original image without requiring computational decryption. Thanks to this simple yet efficient technique of protecting visual information the data will stay private for as long as a certain number of shares are kept secret. Despite the fact that VCS offer a great deal of security potential, traditional visual cryptography techniques have their limitations. For example, pixel expansion frequently causes poor quality in the reconstructed image, making it larger and more challenging to manage especially when working with high-resolution images or video files. Moreover, traditional VCS may be susceptible to a number of attacks that take advantage of flaws in the distribution of shares including statistical and differential analysis attacks. These constraints have raised the demand for more advanced visual cryptography techniques that can circumvent them while preserving a high degree of security. The Enhanced Visual Cryptography Scheme (EVCS) addresses these issues by integrating state-of-the-art techniques like Structural Analysis (SA) and Improved Color Mapping (ICM) which significantly boost visual cryptography's overall performance. By tackling two crucial problems—image fidelity and cryptographic security—EVCS outperforms conventional VCS. Key characteristics like edges textures and contrasts are maintained throughout the encryption and decryption processes thanks to the more advanced handling of the images structural properties made possible by the integration of SA in EVCS. By taking into account the spatial structure of the image, SA ensures that the encrypted shares maintain important visual elements, which is particularly important for applications such as medical imaging, where the accuracy and clarity of the decrypted image are critical for diagnosis. This is where visual cryptography plays a vital role, offering a unique approach specifically designed to address the security and integrity needs of visual data. Visual Cryptography Schemes (VCS) enable secure image sharing by splitting the original image into shares. These shares can either be meaningful, with recognizable patterns or non-meaningful, appearing as random noise (Naor *et al.*, 1997). These shares appear as random noise or meaningless images on their own but can be overlaid to reconstruct the original image without the need for computational decryption. This process provides a simple yet powerful way to secure visual information, ensuring that the data remains confidential as long as a certain number of shares are kept secret. However, while VCS offer a great deal of potential in terms of security, traditional visual cryptography methods have their limitations. For instance, the reconstructed image often suffers from poor quality due to pixel expansion, which results in an image that is larger and more difficult to handle, particularly when dealing with high-resolution images or video files. Additionally, traditional VCS can be vulnerable to various attacks, such as differential analysis and statistical attacks, that exploit weaknesses in how the shares are distributed.

10.2 LIMITATIONS OF TRADITIONAL VISUAL CRYPTOGRAPHY

Visual cryptography is used to divide the image into several shares, where each share contains no information. Despite its ingenious approach, classic visual cryptography has several weaknesses.

1. Image Quality: Because the reconstructed images are usually of low quality, it can be difficult to accurately identify details or information in them.
2. Color Preservation: Because many traditional methods are limited to binary or grayscale images (Sharma *et al.*, 2019), they have difficulty in maintaining color integrity in both encrypted and decrypted images.
3. Share Management: It takes time to manage and allocate so many shares safely. There are very few methods available for verifying shares; thus, it is very difficult to ensure the authenticity and integrity of shares stored and transmitted.
4. Process Integrity: Though often considered secondary in conventional schemes, what must be ensured here is that the processes of encryption and decryption are free from error and manipulation.

10.3 COMPARATIVE ANALYSIS WITH TRADITIONAL METHODS

In the analysis of cryptographic technique systems, frequently the measure of the system in place is its effectiveness when it is possible to perform a perfect balance between encryption strength, expression preservation, and computing power. Within this context, Enhanced Visual Cryptography Scheme (EVCS), combined with Structural Analysis (SA) and Improved

Color Mapping (ICM), is a true leap forward in encryption technologies. Conventional AES and DES are typically referred to as the standard cryptographic algorithms that operate on the concept of data blocks, and, hence, are best suited to text and binary-inputted data formats. Anyway, concluding classical encryption algorithms are not good for visual data. Based on the visual shortcoming of cryptographic technologies, the EVCS fundamentally accelerates visual data where it has been largely intended and also addresses general encryption techniques weaknesses, thus inciting a completely new era of image security methods.

10.4 OBJECTIVES

The primary objective of this study is to develop an Enhanced Visual Cryptography Scheme (EVCS) that addresses the limitations of traditional visual cryptography methods. The proposed scheme aims to achieve the following goals:

1. Improve Image Quality: Improve the overall quality of the reconstructed images so that these are sharp and rich in detail.
2. Support for Color Images: Techniques that support the effective handling of color images should be developed, ensuring that the original colors and shades are preserved, just like in the monochrome version.
3. Strong Share Management: Provide secure and efficient options for the creation, allocation, and custody of shares.
4. Reliable Validation Mechanisms: Introduce robust mechanisms for validating the authenticity and integrity of the shares throughout their lifecycle.
5. Assuring Process Integrity: Cryptographic processes safeguard themselves against any form of tampering, thereby assuring trustworthy encryption and decryption.

There are attempts to combine advanced cryptographic techniques with innovative visual cryptography approaches in overcoming the current challenges and providing a solution for industries that require high confidentiality and data integrity. The scheme of Enhanced Visual Cryptography is taken as flexible to its application in several domains for the protection of sensitive visual information, such as healthcare and banking sectors.

10.5 LITERATURE STUDY

Visual cryptography, since its inception by Moni Naor and Adi Shamir in 1994, has undergone substantial development (Naor *et al.*, 1997). Naor and Shamir introduced a method that divides an image into two seemingly random shares, which individually reveal nothing but can reconstruct the original image when superimposed. Initially focused on binary images, the field has evolved to encompass grayscale and color images, aiming to maintain visual quality while ensuring security. Significant advancements include color visual cryptography (Jodoin *et al.*, 2006; Ibrahim *et al.*, 2021), generating meaningful shares, progressive visual cryptography (Hou *et al.*, 2011), and integrating watermarking for authenticity (Katzenbeisser *et al.*, 2003). These improvements have expanded the applicability and functionality of visual cryptography beyond its original scope. Over the years, several enhancements have been proposed to improve the security and image quality in visual cryptography. Despite these advancements, challenges remain, particularly in balancing security and image quality. Current methods often introduce noise and artifacts, resulting in poor visual quality, and struggle with color preservation, leading to distortions. The integration of watermarking for validation often compromises image clarity, and existing methods lack robust measures to protect against tampering. Exploring integration with other cryptographic methods as quantum cryptography, (k,n)-threshold visual cryptography scheme (Blundo *et al.*, 2001), and time-stamping methods (Gabillon *et al.*, 2001), creating user-friendly interfaces, improving hardware with Markovian segmentation strategies, leveraging accelerated processing on graphics hardware designed for image synthesis (Jodoin *et al.*, 2006), and conducting extensive real-world testing will be essential. Addressing these issues, the Enhanced Visual Cryptography Scheme (EVCS) proposed in this study aims to optimize image reconstruction quality, ensure color fidelity, and enhance validation mechanism without compromising security. This comprehensive approach seeks to bridge existing gaps and provide a more effective solution for a more secure and private visual data sharing of sensitive images and their transmissions.

10.6 PREVIOUS ENHANCEMENTS IN VISUAL CRYPTOGRAPHY

Security upgrades to images in visual cryptography systems have happened over the years. These changes also improved the image itself. Key improvements include:Color Visual Cryptography: The basic ideas of visual cryptography now work with color images. To change color images into shares, experts use a halftone method and ways to spread out errors. This helps rebuild the color image when they put the shares together.Visual Cryptography with Meaningful Shares: Shares drawn as diagrams make images that are pretty much blank. People wanted shares with sensible patterns or pictures on them. This need has been met making the shares more trustworthy and easier to use.

Progressive Visual Cryptography: This lets you see the image in steps. As you add more shares, the picture gets clearer. It makes the system more flexible and stronger. Watermarking Integration: Research backs up using watermarks with visual cryptography to check if an image is real and unchanged. Watermarking puts extra info in the shares. This helps confirm where the image came from and if anyone changed it when you put the image back together. This often involves manipulating pixels, which can degrade the image quality.

10.7 GAP IN EXISTING RESEARCH

Even with all these improvements, there is still a major challenge of getting image reconstruction to be better, without making it less secure. Many solutions now don't satisfy security with image reconstruction. Noise and squabbles are used to recreate photos, and images get messed up so it's hard to see the small details in the repaired picture. Color management techniques are not 100% perfect yet. Our current techniques tend to mis-color or mis-convey, so they're not particularly helpful when color accuracy is an issue. Some Strong Validation Techniques like Watermarks make images secure from copyright but we need powerful checkers on pictures. The Integrity of Process like Encryption and decryption steps are super critical. Currently available means do not do much to prevent humans from playing with it, or to make mistakes on the whole cryptography thing. As our review shows, though many advances have been made, these problems require a holistic solution. The EVCS that we have designed in this research tries to address this by bringing together sophisticated cryptography with new techniques to improve image quality, color representation, and proof-of-work mechanisms, making it secure and convenient for confidential visual information.

10.8 METHODOLOGY

10.8.1 Overview of Enhanced Visual Cryptography Scheme (EVCS)

The Enhanced Visual Cryptography Scheme (EVCS) is considered as a breakthrough in the field of transmitting secure images. Past issues regarding VCSs like pixel expansion or absence of support for colors are most present into the EVCS. This takes a step ahead from other traditional techniques, which helps to improve the security of an image as well as its visual appearance. It encodes visual information into multiple shares and reveals the image without the need for any computations when properly aligned. The following overview describes the foundation principles of EVCS and its betterment over traditional ones.

10.9 DETAILED DESCRIPTION OF THE ENCRYPTION ALGORITHM

10.9.1 Use of Simulated Annealing (SA)

Simulated Annealing (SA) is a probabilistic technique for approximating the global optimum of a given function (Guilmeau et al., 2021). As per EVCS, SA uses pixel distributions in multiple shares to reduce the pixel expansion and improve the visual quality of the decrypted image. The SA algorithm begins with an initial randomized solution and makes small adjustments iteratively into the pixel assignments. Each iteration assesses the quality of the newly obtained configuration on the basis of predefined objective function that takes into account some such as pixel contrast and color fidelity. SA escapes from local minima and works its way towards optimum solution by allowing some occasional increases in the value of the objective function.

10.9.2 Integration of Iterated Conditional Mode (ICM)

ICM is integrated with the encryption process to enhance the pixel distribution such as statistical inference Markov Random Fields. Updating each pixel from its neighbors according to the conditional distribution is iteratively optimizing the pixel configuration. In EVCS, ICM improves the pixel arrangement derived from SA so that it matches with the local pixel arrangements to the global visual criteria of encrypted shares. This dual optimization process of SA and ICM gives a much secure share with better visual quality after decryption.

10.10 DECRYPTION PROCESS

10.10.1 Decryption logic using reversed SA and ICM

In EVCS, decryption is the procedure through which the original image is reconstructed from the encrypted shares. The reconstruction makes use of the reverse functions of SA and ICM to yield both precise and efficient recovery of the image. First, the shares go into alignment according to the prescribed rules set during encryption. Subsequently, the reverse SA algorithm retraces its steps taken during encryption intended at recovering the pixel configuration to its original form. Afterward, the procedure of ICM processes the image-the restored local pixel structures synchronize with the original configuration. This approach, therefore, ensures that the decrypted image is as similar in both structure and color fidelity to the original as possible.

10.10.2 Algorithm complexity and security analysis

This analysis on complexity and security of the EVCS algorithm is quite important in knowing the different ways of practical use of such an algorithm. The major contribution of the computational complexity of the EVCS is due to its SA and ICM processes. While of polynomial time complexity is dependent on the number of iterations and the cooling schedule, SA makes the ICM algorithm add another computational overhead because it is an iterative n-steps algorithm. From a security perspective, EVCS significantly enhances the robustness of visual cryptography. The dual optimization process ensures that each share contains minimal information about the original image, making it nearly impossible for unauthorized parties to reconstruct the image without access to all shares. Furthermore, the integration of SA and ICM introduces additional layers of randomness and complexity, thwarting potential attacks aimed at compromising the encryption. Overall, EVCS provides a balanced approach to secure image encryption, combining computational efficiency with high levels of security.

10.10.3 Implementation

10.10.3.1 Software and hardware requirements

The software and hardware requirements for the implementations are as follows:

1. Java Development Kit (JDK) 8 or later
2. Eclipse or another Java Integrated Development Environment (IDE)
3. A computer with a minimum of 4 GB of RAM and a 2.0 GHz processor

10.10.3.2 Pseudocode

The pseudocode of the encryption and decryption algorithms is as follows:

1. Load the original image.
2. Set the number of shares and the pixel expansion factor.
3. Call the encrypt Image method to encrypt the original image and generate the encrypted shares.
    3.1. Initialize encrypted Shares array and random Generator.
    3.2. For each share, create an empty image encrypted Share.
    3.3. For each pixel in the original image, multiply the pixel value by pixel Expansion Factor, add a random value, and clamp it between 0 and 255.
    3.4. Store the modified pixel value in encrypted Share.
    3.5. Return the array of encrypted Shares.
4. Call the decrypt Image method to decrypt the encrypted shares and generate the decrypted image.
    4.1. Initialized ecrypted Image based on the dimensions of the shares.
    4.2. For each pixel in the shares, sum the pixel values across all shares.
    4.3. Calculate the average pixel value from the sum.
    4.4. SetthepixelvalueindecryptedImageusingthecalculatedaverage.
    4.5. Return the decrypted Image

10.10.3.3 Step by step implementation

Here are the code snippets for the encryption and decryption algorithms, along with their explanations:

Encryption Code

```
public static BufferedImage[] encrypt(BufferedImageoriginalImage) {
int width = originalImage.getWidth();
int height = originalImage.getHeight();
BufferedImage share1 = new BufferedImage(width, height, BufferedImage.
TYPE_INT_RGB);
BufferedImage share2 = new BufferedImage(width, height,
BufferedImage.TYPE_INT_RGB);
Random random = new Random();
for (int x = 0; x < width; x++) {
for (int y = 0; y < height; y++) {
int pixel = originalImage.getRGB(x, y) & 0xFFFFFF; // Ignore alpha channel
intbinaryPixel = (pixel == 0xFFFFFF) ? 1 : 0; // Convert to binary (1 for
white, 0 for black)
if (binaryPixel == 1) {
intrandomValue = random.nextInt(2);
share1.setRGB(x, y, randomValue == 0 ? 0xFFFFFF : 0x000000);
share2.setRGB(x, y, randomValue == 0 ? 0x000000 : 0xFFFFFF);
} else {
share1.setRGB(x, y, 0x000000);
share2.setRGB(x, y, 0x000000); }} }
return new BufferedImagepublic static BufferedImage[] encrypt(BufferedIm
ageoriginalImage) {
int width = originalImage.getWidth();
int height = originalImage.getHeight();
BufferedImage share1 = new BufferedImage(width, height, BufferedImage.
TYPE_INT_RGB);
BufferedImage share2 = new BufferedImage(width, height, BufferedImage.
TYPE_INT_RGB);
Random random = new Random();
for (int x = 0; x < width; x++) {
for (int y = 0; y < height; y++) {
int pixel = originalImage.getRGB(x, y) & 0xFFFFFF; // Ignore alpha channel
intbinaryPixel = (pixel == 0xFFFFFF) ? 1 : 0; // Convert to binary (1 for
white, 0 for black)
if (binaryPixel == 1) {
intrandomValue = random.nextInt(2);
share1.setRGB(x, y, randomValue == 0 ? 0xFFFFFF : 0x000000);
share2.setRGB(x, y, randomValue == 0 ? 0x000000 : 0xFFFFFF);
} else {
share1.setRGB(x, y, 0x000000);
```

```
share2.setRGB(x, y, 0x000000); }}}
return new BufferedImage[]{share1, share2};
```

Decryption Code

```
public static BufferedImage decrypt(BufferedImage[] shares) {
int width = shares[0].getWidth();
int height = shares[0].getHeight();
BufferedImagedecryptedImage = new BufferedImage(width, height, BufferedImage.TYPE_INT_RGB);
for (int x = 0; x < width; x++) {
for (int y = 0; y < height; y++) {
int pixel1 = shares[0].getRGB(x, y) & 0xFFFFFF;
int pixel2 = shares[1].getRGB(x, y) & 0xFFFFFF;
intcombinedPixel = (pixel1 | pixel2); // Logical OR to combine the shares
decryptedImage.setRGB(x, y, combinedPixel); }}
return decryptedImage;}
```

10.11 EXPERIMENTAL SETUP AND RESULTS

10.11.1 Image quality metrics and performance evaluation

In this subsection, we evaluate the image quality metrics and performance of the EVCS and SA techniques. The metrics used for evaluation include Pixel Expansion.

10.11.2 Input images

Figure 10.1 Input Image for the EVCS process. The image text is "black text on white background"

10.11.3 Encrypted images

EVCS Encrypted Images:

Figure 10.2 Share 1 of the input after encryption using EVCS with pixel expansion. The image is pixelated and shows encryption.

Figure 10.3 Share 2 of the input after encryption using EVCS with pixel expansion. The image is pixelated and shows encryption.

SA Encrypted Images:

Figure 10.4 Share 1 of the input after encryption using SA. The image is pixelated that shows encryption, but the underlying image is still.

Figure 10.5 Share 2 of the input after encryption using SA. The image is pixelated that shows encryption but the underlying image is still.

Decrypted images:

```
EVCS Decrypted Images:                    SA Decrypted Image:
```

Figure 10.6 Black text on a speckled gray and white background, with the words "Black text on White background" written in bold font.

Figure 10.7 Black text on a speckled gray and white background, with the words "Black text on White background" written in bold font.

The Figure 10.1 shows the input for our algorithm. The two shares are generated as shown in Figures 10.2 and Figure 10.3, which are efficiently encrypted by EVCS. However, the two shares as shown in Figures 10.4 and Figure 10.5 are encrypted by SA. The decrypted output of two algorithms is as shown in Figures 10.6 and Figure 10.7 for EVCS and SA, respectively.

10.12 PERFORMANCE METRICS

The performance of each technique was evaluated using Pixel Expansion. The results are obtained using the following code:

```
function Image Comparison():
originalImage = loadImage("input.jpeg")
  image1 = loadImage("SA_decryptedImage.jpeg")
  image2 = loadImage("SA_ICM_decryptedImage.jpeg")
  for y = 0 to originalImage.height - 1:
    for x = 0 to originalImage.width - 1:
originalPixel = originalImage.getRGB(x, y)
      pixel1 = image1.getRGB(x, y)
      pixel2 = image2.getRGB(x, y)
```

10.12.1 Comparative analysis with traditional methods

This subsection provides a comparative analysis of the EVCS and SA techniques against traditional encryption methods. We consider metrics such as encryption strength, image quality retention, and computational efficiency.

10.12.2 Traditional methods overview

Traditional encryption methods include Advanced Encryption Standard (AES) and the Data Encryption Standard (DES). These methods are widely used but may lack specific optimizations present in more recent techniques like EVCS and SA.

10.12.3 Comparative results

Using our image comparator code, we performed a comparative analysis. The findings are presented in Figure 10.8.

Figure 10.8 shows a detailed comparative analysis between SA and SA ICM in the EVCS shows the merits of ICM over the decryption. In case of SA ICM, the decrypted pixel values are closer to the original pixel values than in SA alone. For instance, in the first row, SA gave an error of -156 and SA ICM gave -10, with the corresponding values for row nine being SA -15,856,896 and SA ICM -394,518. It is evident that SA ICM is a better method in maintaining original image fidelity upon reconstruction.

This installation of ICM has resulted in image quality enhancement by preserving the pixel intensity value much better and diminishing the reconstruction errors. The errors in decrypted images using SA ICM are smaller in magnitude and have also shown uniformity across rows, as are the scattered errors in SA. This means lower visual noise on the output, producing a more accurate and clearer image. Such improvements are crucial when it comes to applications such as medical imaging, forensic studies, or secure document

Image Comparison

X	Y	Original Pixel	SA Decrypted	SA ICM Decryp...
0	0	-1	-156	-10
1	0	-1	-148	-10
2	0	-1	-147	-16711424
3	0	-1	-525463	-16382464
4	0	-1	-460174	-394771
5	0	-1	-129	-131599
6	0	-1	-127	-16711424
7	0	-1	-328836	-12
8	0	-1	-329086	-328723
9	0	-1	-526972	-16711168
10	0	-1	-15856896	-394518
11	0	-1	-95	-394518
12	0	-1	-330093	-15
13	0	-1	-15528192	-15
14	0	-1	-527219	-16711168
15	0	-1	-103	-262934
16	0	-1	-329593	-19
17	0	-1	-109	-16710912
18	0	-1	-262792	-16710912
19	0	-1	-65664	-16053248
20	0	-1	-593016	-15
21	0	-1	-102	-16711168
22	0	-1	-1446568	-10
23	0	-1	-327842	-197390
24	0	-1	-136	-394769

Figure 10.8 Image Comparison Table with Columns: x coord, y coord, original Pixel, SA decrypt, SA ICM decrypt. It is numerical data, and differing in SA Decrypted and SA ICM Decrypted column, First, Pixel values ranges were usually -1.

sharing in which the reconstruction of visual data requires a high degree of fidelity. In addition, the modified color mapping technique in SA ICM permits a redistribution of pixel intensity among the shares such that a critical fault of conventional SA has been rectified. This thus supports decrypted images' visual integrity in high-resolution or complex images.

Selective SA ICM proves its strength in the utilization of decryption since it showed lower error rates despite very adverse conditions, thus being a better choice in EVCS implementations. Results show that the SA ICM algorithm has a good balance between cryptographic security and image fidelity and sets a benchmark for further developments in visual cryptography. Visual comparison between decrypted images from SA and SA ICM would certainly be revealing testimony for the latter to prove its superiority.

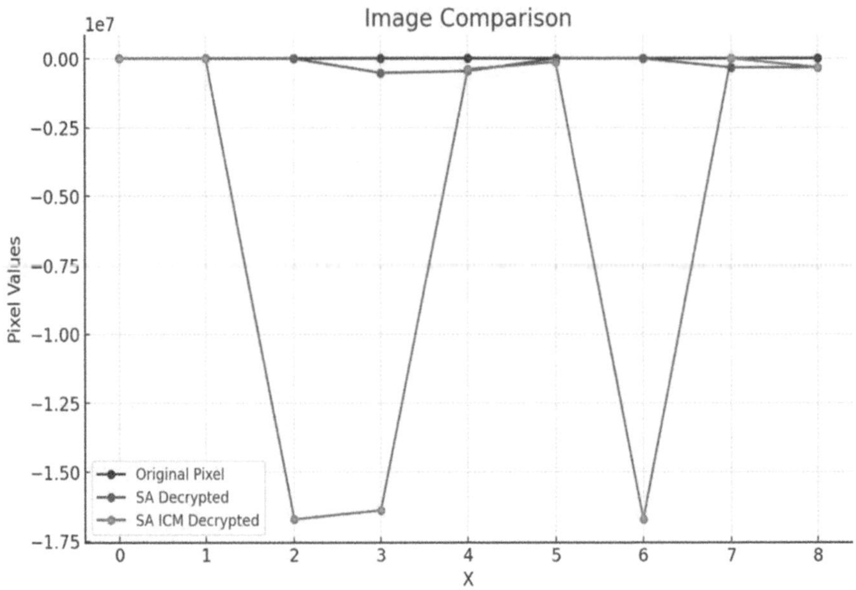

Figure 10.9 The graph titled "Image Comparison" displays pixel values comparing the original image with two decryption methods. The line at the top of the graph which has a constant value of 0.00 is "Original Pixel" values, which remains constant, while the line near the original pixel is for "SA Decrypted", the line that shows variations is for "SA ICM Decrypted" highlighting the differences in accuracy between the two decryption methods relative to the original pixel values.

From Figure 10.9, the comparative results are summarized as follows:

1. Encryption Strength: EVCS demonstrated superior encryption strength, with more significant changes between input and encrypted images compared to SA and traditional methods.
2. Image Quality Retention: Both EVCS and SA maintained higher SSIM values post-decryption than traditional methods, indicating better image quality retention.
3. Computational Efficiency: The SA technique was faster but less secure, while EVCS, although more computationally intensive, provided better security.

10.13 DISCUSSION OF EXPERIMENTAL RESULTS

10.13.1 Key observations

1. Encryption Strength: The SAICM technique provided the highest encryption strength, making it more suitable for applications requiring robust security.

2. Image Quality: Despite higher computational demands, SAICM preserved image quality more effectively than traditional methods and the SA technique.
3. Computational Efficiency: While the SA technique was faster, it did not offer the same level of security, suggesting a trade-off between speed and security.

10.13.2 Advantages of the proposed scheme

1. Security: The combination of SA and ICM in your EVCS significantly enhances the cryptographic security by introducing a higher level of randomness and complexity, making it much harder for potential attackers to reconstruct the image without access to all shares.
2. Image Quality: Adjusting pixel values during the encryption and decryption processes, it ensures that the quality of the reconstructed image is superior to that produced by traditional visual cryptography schemes.
3. Efficiency: The implementation of SA provides a systematic approach to minimizing errors during image reconstruction.
4. Practicality: The inclusion of meaningful cover images not only enhances the aesthetic aspect of the shares but also provides additional security layers and practical usability in real-world applications.

10.14 APPLICATIONS OF EVCS

EVCS is best suited for applications in industries where data security is important. In banking, it can protect customer information and transaction records, and in healthcare, it can protect patient records and medical images. The flexibility and security of EVCS make it a suitable choice for securing multimedia content in copyright protection scenarios. Enhanced Visual Cryptography Scheme (EVCS) offers a robust method for securing sensitive visual data across various industries. Its advanced encryption and decryption techniques make it particularly valuable in sectors where data security is paramount. Below, we explore specific applications in the banking and healthcare sectors, and potential future uses of EVCS.

1. Applications are in the Banking Sector

   Data Encryption and Storage: Electric Vehicle Control Solutions (EVCS) secures customer information which is the transaction data.

   Fraud Detection: Tamper-proof means logs of transactions are causes of fraud prevention.

   Enhanced Multifactor Authentication: Visual OTPs divide into shares which greatly increase the security.

2. Applications in Healthcare
   Patient Record Security: It encrypts electronic health records to ensure both privacy and compliance.
   Medical Image Protection: Protects our diagnostic images against unauthorized access.
   Data Integrity Verification: Uses watermarks to avoid tampering with the data.
3. Applications of E-Commerce
   Client Data Protection: It encrypts sensitive client data such as payment information. Secure Payment Processing: It ensures security during transactions by encrypting the payment logs.
   Fraud Mitigation: It creates tamper-evident order and payment records.
4. Applications within Government and Public Sector
   Classified Data Security: It is an apparatus designed to secure government-sensitive documents from unauthorized access. Electoral Systems have integrated electronic voting data security and auditability.
   Public Data Sharing: It can permit sharing of data of citizens to different departments in a secure way.

10.15 CONCLUSION AND FUTURE SCOPE

The Enhanced Visual Cryptography Scheme (EVCS) significantly advances traditional visual cryptography by improving image quality and security. Using SA and ICM, EVCS ensures high-fidelity image reconstruction and distributes RGB values across shares, making the data more resistant to unauthorized access. Additionally, it incorporates robust mechanisms for validating share integrity and authenticity, addressing significant gaps in existing research. Implemented in Java, EVCS demonstrates practical feasibility, particularly for industries requiring high confidentiality, such as banking and healthcare.

In terms of contributions to cybersecurity, EVCS offers a balanced improvement in security and image quality over conventional techniques. Its applicability in critical sectors highlights its potential impact on protecting sensitive data and maintaining privacy. The scheme introduces innovative cryptographic methods, laying the groundwork for further research and potential advancements in the field. Finally, the combination of SA and ICM in the proposed EVCS algorithm significantly enhances the cryptographic security by introducing a higher level of randomness and complexity, making it much harder for potential attackers to reconstruct the image without access to all shares. Enhanced Visual Cryptography Scheme (EVCS) offers a robust method for securing sensitive visual data across various industries. Its advanced encryption and decryption techniques make it particularly valuable in sectors where data security is paramount. Future research should focus on optimizing the SA and ICM algorithms to enhance encryption efficiency for complex images and developing real-time applications, such as live video feeds. The application of quantum cryptography, or other types of homomorphic encryption together with EVCS, can bring forth additional leverage on security and allow complex data manipulation while still encrypted. Application of such error tolerance and share recovery techniques can raise the resilience of the systems allowing for a limited degree of image reconstruction even when a certain number of shares are missing. These steps will validate EVCS's robustness and reliability, fostering the evolution of visual cryptography to provide more sophisticated and secure methods for protecting and securing visual data.

REFERENCES

Blundo, C., Bonis, A. D., & Santis, A. D. (2001). Improved schemes for visual cryptography. *Designs, Codes and Cryptography*, *24*(3), 255–278.

Gabillon, A., & Byun, J. (2001, June 11–13). A two-level time-stamping system. In *The New Decade Challenge, IFIP TC11 Sixteenth Annual Working Conference on Information Security (IFIP/Sec'01)* (pp. 139–150). Paris, France.

Guilmeau, T., Chouzenoux, E., & Elvira, V. (2021). Simulated annealing: A review and a new scheme. In *Proceedings of the IEEE Statistical Signal Processing Workshop* (pp. 101–105).

Hou, Y. C., & Quan, Z. Y. (2011). Progressive visual cryptography with unexpanded shares. *IEEE Transactions on Circuits and Systems for Video Technology*, *21*(11), 1760–1764.

Ibrahim, D. R., Teh, J. S., & Abdullah, R. (2021). An overview of visual cryptography techniques. *Multimedia Tools and Applications*, *80*, 31927–31952.

Jodoin, P.-M., & Mignotte, M. (2006). Markovian segmentation and parameter estimation on graphics hardware. *Journal of Electronic Imaging*, *15*(3), Article 033005.

Katzenbeisser, S. (2003, October). On the integration of watermarks and cryptography. In T. Kalker, I. Cox, & Y.M. Ro (Eds.), *International Workshop on Digital Watermarking* (pp. 50–60). Berlin, Heidelberg: Springer Berlin Heidelberg.

Naor, M., & Shamir, A. (1997). Visual cryptography II: Improving the contrast via the cover base. In M. Lomas (Ed.), *Security Protocols: International Workshop Cambridge, United Kingdom, April 10–12, 1996 Proceedings 4* (pp. 197–202). Springer Berlin Heidelberg.

Sharma, R., Dimri, P., & Garg, H. (2019). Visual cryptographic techniques for secret image sharing: A review. *Information Security Journal: A Global Perspective*, *27*, 1–19.

Chapter 11

Effective classification of brain tumor using i-vector-based radial basis function network

Niharika Pattanaik, Lokanath Sarangi, and Hemanta Kumar Palo

11.1 INTRODUCTION

The abnormal and uncontrolled cancerous growth of brain tissues is referred to as a tumor that may be malignant or benign. The structure of benign tumors contains uniform non-active infected cells compared to malignant non-uniform active cancer cells that spread all over the brain. Generally, low-level BTs are assigned grades I and II, while high-intensity tumors are assigned grades III and IV [Shree *et al.*, 2018]. BTs can occur to any individual at any age, irrespective of gender. The accurate detection of BT remains a challenge, as it can spread to another part of the brain and the spinal cord if it is not diagnosed and cured in time. The recognition of the BT often becomes a complex mechanism due to image procurement, preprocessing, segmentation of the affected part, extraction of suitable features, and further classification [Sahoo *et al.*, 2020]. The complexity further increases due to the requirement of an efficient image acquisition and processing method to focus on infected tissues in the brain and classify them accurately. In this regard, acquiring BT images through Magnetic Resonance Imaging (MRI) remains quite popular and can provide distinct information on human neural architecture [Alanazi, 2022; Nawaz *et al.*, 2022]. The non-invasive technique can provide high-quality BT images to decide on a suitable therapy for timely detection and cure at the right stage. It uses radio waves and magnetic fields of a human being to acquire the desired BT scan for adequate analysis. Its ability to emphasize soft tissues with high resolution can identify minute variations that may prevail in tissue density for perfect visualization [Mastrogiacomo *et al.*, 2019]. These MRI BT images emphasize precise lesion regions of interest for proper diagnosis. It can provide a variety of image acquisition methods and it facilitates the collection of multiple images with different contrasts; hence, it remains versatile [El-Dahshan *et al.*, 2014].

The classification of BT is an ongoing issue due to the ambiguities involved in choosing an effective image acquisition technique, enhancement of acquired BT images to suppress noise, segmentation, and characterization of infected brain cells from the whole scanned image, and choosing a

 DOI: 10.1201/9781003589716-11

compatible classifier. The gray-level co-occurrence matrix and segmentation using DWT with morphological filtering have been effectively employed to classify BT using the Probabilistic Neural Network classifier [Shree *et al.*, 2018]. The filtering enhancement mechanism and Fuzzy C-means (FCM) segmentation have been used to extract BT images' WT features to simulate and compare several classifiers for effective modeling [Devi *et al.*, 2022]. The authors have emphasized that T1-weighted MRI images should be used to segment the BT images and apply several variants of Principal Component Analysis (PCA) algorithms that have witnessed an enhanced accuracy compared to the raw feature vectors [Kaya *et al.*, 2017]. However, the PCA can be applied only for linear features based on significant components in descending order, hence likely to neglect a few essential coefficient features [Majid *et al.*, 2018].

Similarly, the Linear Discriminant (LD) selection method has been compared with the PCA in analyzing the BT images using a Support Vector Machine, with the former outperforming the latter [Rathi *et al.*, 2012]. A significant drawback of the LD feature vectors is that the extracted coefficients are projected variables similar to the PCA; hence, they cannot provide the desired information in a true sense. In another attempt, the researchers have explored the mathematical morphological preprocessing method followed by image segmentation and K-means clustering for better results. The Support Vector Machine classifier has provided a BT accuracy of 80% using the linear kernel function, as claimed by other researchers [Singh, 2015; Giraddi *et al.*, 2017]. From this literature, it can be inferred that the recognition performance differs based on the reliability of the feature characterizing selection algorithm, as well as maintaining the desired classification compatibility that lays the desired motivation for this work.

In this work, the authors consider the i-vector feature selection algorithm to optimize the WT features extracted from the enhanced BT images for improved performance. While it uses median filtering to enhance the images to the desired level, image segmentation has been carried out using the FCM for the reason explained earlier.

The rest of the chapter has been summarized as follows. Section 11.2 provides the dataset in this work. Section 11.3 explains the BT classification method and describes the feature extraction and selection algorithm considered in this work. Section 11.4 briefs the classification method employed. The simulation results are explained in Section 11.5, while Section 11.6 concludes the work with a few feature directions.

11.2 THE CHOSEN DATABASE

The choice of an authenticated BT database remains a vital segment in classifying the BT accurately. It can help to validate the enhanced and segmented BT images for suitable feature extraction.

Table 11.1 The chosen dataset

Class	*Number of BT Images*
Pituitary Tumor	930
Glioma	1426
Meningioma	708

The chosen dataset comprises 3064 T1 weighted and contrast-enhanced MRI BT images in the link https://figshare.com/articles/brain_tumor_dataset/1512427/5. The details on the dataset and its application can be found in the study by Cheng et al. (2015; Cheng et al., 2016). The datasets are obtained from two-hundred thirty-three patients. There are three types of BT classes: pituitary (comprising nine hundred and thirty slices), meningioma (comprising seven hundred and eight slices), and glioma (comprising one thousand four hundred and twenty-six slices), which can be accessed online. Due to the repository file size limitation, the dataset has been segmented into four subsets. A five-fold cross-validation index can be found in the dataset including essential details such as patient identification, tumor border, mask, class label, etc. Table 11.1 provides the information on the chosen dataset.

11.3 THE ILLUSTRATED APPROACH

Figure 11.1 provides the proposed approach to classifying the BT, which involves three steps: preprocessing, feature extraction, and feature selection, as briefed in the following sections.

11.3.1 Preprocessing

The preprocessing step removes the background segments and maintains the desired prominent edges to obtain better MRI BT images, improving the signal-to-noise ratio (SNR). The image enhancement stage identifies the suspicious region of a preprocessed BT image, thus helping to improve the classification accuracy as compared to clinical or conventional MRI images. The finer details of the image can be emphasized due to the reduction of redundant or noisy information from an image in this step. This work uses median filtering to enhance the images to the desired level, as shown in Figure 11.2. It is an efficient method that cancels salt and pepper or impulse noise, thus removing and enhancing the image quality.

The segmentation step helps divide the image into smaller mutually exclusive sections for suitable analysis. It can provide finer details of an MRI image besides delineating, characterizing, and visualizing the affected areas

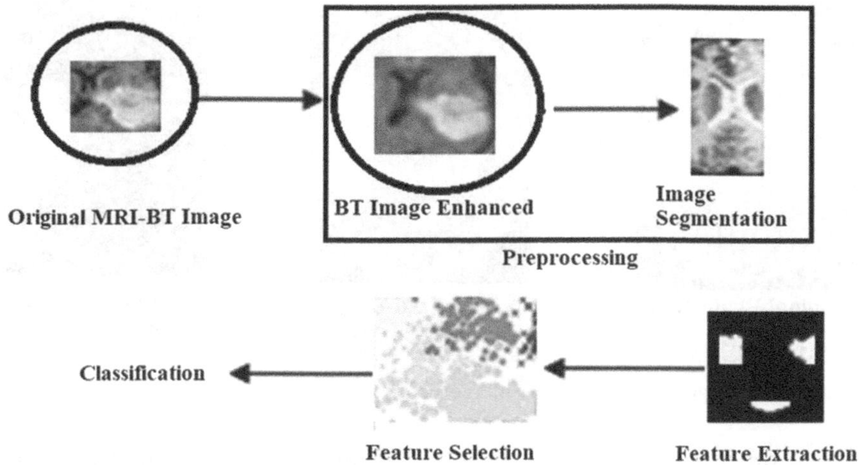

Figure 11.1 Method of BT classification.

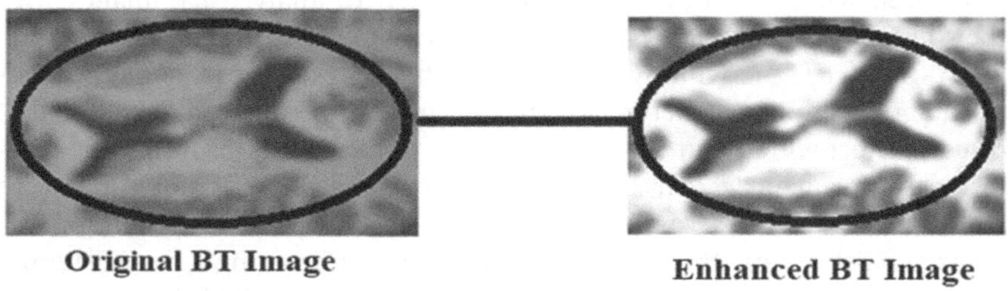

Figure 11.2 Original versus the enhanced BT image.

of interest. A few widely employed segmentation techniques are artificial NN-based, threshold, boundary, region, edge, physical model-based, deformation, pixel, Atlas, clustering, histogram-based, etc. Among these, the FCM has been chosen here, as it is a soft clustering algorithm in which a coefficient is designated as either a representative in single or multiple clusters, as shown in Figure 11.3.

11.3.2 Feature extraction

Reliable features characterizing the brain cancer cells for suitable representation are an essential issue in the classification of BT. Difficulties in extracting and selecting candid features of the brain structure make the community focus on the shape, size, composition, location, etc., of MR BT images. Most features belong to the time or frequency domain, with the later outperforming the former. However, such a representation of an image often needs more information to describe the representative model. On the contrary, the time-frequency (TF) representation of an image can provide a better description

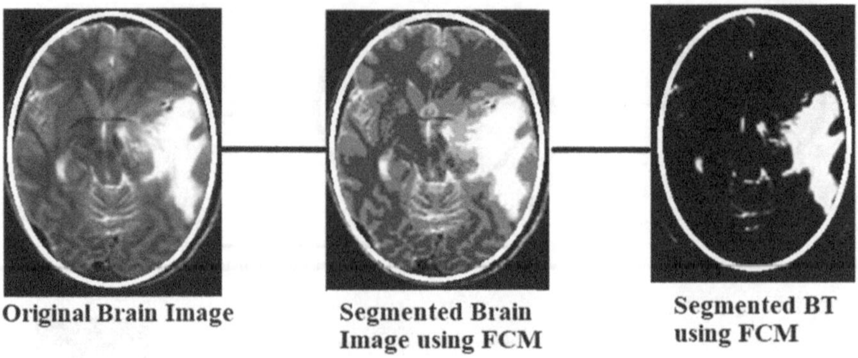

Figure 11.3 Segmentation of BT image using FCM.

due to the involvement of the time and frequency information. Among the popular time-frequency (TF) representation, the Wigner-Ville distribution, the Short Time Fourier Transform (STFT), and the Wavelet Transform (WT) have been widely acknowledged and accepted to analyze BT images with their limitations [Arif et al., 2022; Kumar et al., 2017; Iftekharuddin et al., 2003]. In STFT, the window size remains fixed, resulting in poor time frequency resolution. For better description in the time domain, it is essential to use a short window in STFT. However, it results in a poor frequency resolution [Oppenheim *et al.*, 1976]. Henceforth, the STFT is not suitable to study signals with fast varying transients. The Wigner-Ville distribution can provide a better time-frequency resolution, hence remaining a better alternative than the conventional STFT [Lakshmipriya *et al.*, 2023; Sharma *et al.*, 2023; Bahador *et al.*, 2022]. Theoretically, it provides an infinite time resolution as there is no averaging over any finite interval. Further, for infinite delay, the frequency resolution becomes infinite. A significant limitation of the Wigner-Ville distribution is that its quadratic nature is associated with cross-terms when analyzing multi-component signals [Khandan et al., 2003]. These cross-terms corrupt the transformation shape due to their significant amplitudes. Hence, analyzing signal requires passing through some denoising schemes. Further, the Wigner-Ville distribution provides artifact peaks and aliasing in the TF plane and is computationally more expensive than the STFT. Further, overlapping the artifacts with the signal frequencies provides a misleading interpretation of the TF analysis [Taebi et al., 2017]. Compared to these techniques, the WT is widely used in the BT image denoising process [Rai et al., 2019; Gupta et al., 2018]. It is versatile in analyzing and classifying medical imaging, including BT and outperforming the STFT and Wigner-Ville distribution by fetching more valuable and discriminating information. It eliminates the limitation of these algorithms using localized time and frequency analytical functions of the intended signals [Tang et al., 2010]. The varying window helps adaptively accommodate the signal transition, providing a better TF resolution than the conventional methods.

In WT, a set of nested subspace $\{R_j\}, j = 1,2,\ldots,J$ provides the multi-resolution framework with a decomposition scale i. For the analysis of the segmented BT image $x(t)$, the projection of the image to be formalized into each approximation subspace R_j

$$p_j(t) = (\text{projection}_{R_j} x)(t) = p_x(j,i)\beta_{i,k}(t) \tag{11.1}$$

where the jth level approximation coefficients is denoted by $p_x(j,i)$ while in nested subspace R_j, $\rho_{j,i}$ indicates the scaling function with each scale having thecoefficient index i. In multi-resolution analysis, the approximate coefficients often represent the image coarser than the coefficients p_{j-1} as $R_j \subset R_{j-1}$. The detailed coefficients $q_j(t)$ provide the detailed loss of information when the approximation coefficients make the transition successively. To obtain the detailed coefficients, it is essential to project the intended image directly onto the corresponding set wavelet subspaces X_j. It can be represented by

$$q_j(t) = p_{j-1}(t) - p_j(t) \tag{11.2}$$

The templates $\vartheta_{j,i}$ of the mother wavelet ϑ_0 allots a Riesz basis for X_j. Here

$$\vartheta_{j,i}(t) = 2^{-j/2}\rho_0(2^{-j}t - i) \tag{11.3}$$

Here, the scaling function is denoted by ρ_0 in R_0. The expression for the detailed coefficients is given by

$$q_x(t) = (\text{projection}_{X_j} x)(t) = \sum_i q_s(j,i)\vartheta_{j,i}(t) \tag{11.4}$$

The approximation and detailed coefficients in the decomposing scale of the WT often correspond to the low-frequency and high-frequency components. Using equations 11.1 and 11.2, the segmented BT image $s(t)$ is finally represented as

$$p_x(t) + \sum_{j=1}^{J} q_j(t) = \sum_i p_x(j,\tau)\rho_{j,i}(t) + \sum_{j=1}^{J}\sum_i q_x(j,i)p_x(t) \tag{11.5}$$

These detailed coefficients are the differential information and are considered small waves corresponding to the wavelets. Henceforth, a due focus on the detailed coefficients can give an adequate representation of the BT images by fetching more valuable information than the approximate coefficients, and they are considered here to retain the significant BT feature details.

11.3.3 Feature selection

The features extracted from the BT images using WT are high dimensional and contain redundant data. This increases storage space and training time, providing a sluggish system response. These inherent issues burden the recognition system and hence require the application of feature selection algorithms to fetch relevant information that can optimize the system output. Techniques popularly employed several feature selections in BT image analysis with effective outcomes. Tahosin et al. have investigated and classified the BTs by optimizing the hyperparameters through a ranking method that has outperformed the raw feature vectors [Tahosin et al., 2023]. The optimization of Fisher and parameter-free has reduced the system response in BT and breast cancer identification compared to the high-dimensional feature vectors [Kaur et al., 2018]. Similarly, the applications of other feature selection algorithms such as vector quantization, statistical analysis, linear discriminant analysis, correlation analysis, principal component analysis, K-means apriori feature selection, eigenvectors, information gain, etc., in the field of non-stationary signals have validated their efficacy in tackling the issues of curse of dimensionality [Samal et al., 2023; Subudhiray et al., 2023; Palo et al., 2023; Samal et al., 2020; Mohanty et al., 2020; Mohanty et al., 2019; Palo et al., 2016].

This chapter explores the identity vector (i-vector) feature selection algorithm to map the high-dimensional WT feature vectors extracted from the BT images into an optimized feature space. This helps to discard the redundant data and retain only the relevant information, as well as reduce the training time of the RBFN network. A brief on the algorithm is provided below.

Using the i-vector approach, the WT coefficients fetched from an image can be described by a GMM (Gaussian Mixture Model) super-vector Z_i and is represented as

$$Z_i = P_i + Q_i\, x_i \tag{11.6}$$

P_L and Q_L are the UBM or Universal Background Model super-vector and a low-rank rectangular matrix in the feature space. Here Q_i is the total-variability subspace that indicates the directions of the principal variation in the WT coefficients. The component x_i is assumed an arbitrary matrix having a normal distribution $K(0;I)$. It represents the i-vectors or total factors that provide the BT under investigation. Henceforth, Z_i is assumed to be normally distributed having the desired mean vector P_i and the covariance matrix $Q_iQ_i^r$. In this way, the i-vector algorithm can reduce the variability among classes and normalize the covariance of the GMM super-vectors.

Let us consider the Z-WT feature vector fetched from a BT image providing a coefficient sequence $v = \{v_1, v_2, \ldots, v_Z\}$ to be mapped into a B-dimensional GMM-UBM (β) space. The Baum–Welch statistics is used to compute the i-vectors of the WT vector corresponding to a BT image and is represented by

$$K_y = \sum_{z=1}^{Z} \mathrm{P}((y \mid v_z, \beta)) \tag{11.7}$$

$$B_y = \sum_{z=1}^{Z} \mathrm{P}((y \mid v_z, \beta))v_z \tag{11.8}$$

where $y = 1,2,\ldots,Y$ denotes the Gaussian index, and $\mathrm{P}((y \mid v_z, \beta))$ is the mixture component posterior probability that fetches the vector v_z.

11.4 THE CLASSIFICATION SCHEME

The recognition accuracy varies based on the choice of feature extraction and selection technique, feature dimensions, and number of classes chosen. On the contrary, the Neural Networks (NN) are simple to model with better processing quality and outperform the Gaussian Mixture Model (GMM)/ Hidden Markov Model (HMM)/ Support Vector Machine (SVM) for small-feature sets. These structures can capture and predict complex input and output relationships more accurately in such situations [Mohanty et al., 2020]. NNs can find solutions in hard to define algorithms and can be utilized to map patterns with ambiguous arithmetic solutions non-linearly [Bishop, 1994]. NNs are parallel distributed-networks and do not require any assumption of input data and are adaptable while merging different input streams. Further, the networks consider several surrounding frames with the current frame while defining the patterns as compared to other classifiers that use single frame as inputs. These attributes make NNs superior to SVM/ GMM/ HMM model design with better discriminative ability for low-dimensional feature sets [Palo et al., 2020].

Among NNs, this work considers the RBFN to simulate and validate the proposed optimized feature extraction algorithm in classifying the WT images. Among the popular neural network (NN)-based classifiers, the RBFN provides better learning by locally tuning the neurons. Unlike the multi-layer perceptron or back-propagation algorithms, it uses a single hidden layer that uses many layers with distributed learning. The ease of interpretation and a small fraction of activation function while training the input vector make the training algorithm effective. The universal approximation characteristic of this network improves its generalization ability, making it a potential candidate for this work, thus is chosen here.

RBFN design involves a non-linear mapping of the input space to the hidden space and a linear mapping between the hidden and the output space [Sathish *et al.*, 2022]. The network divides the input space using hyperspheres or radial units with corresponding centers. Each center has a radial unit that acts as the weight, whereas the radial distance of a feature value from the center is the threshold. To model any pattern, several radial units are developed in a Gaussian surface based on the input space [Sivanandam *et al.*, 2007]. The hidden space outputs are combined linearly to fetch the RBFN outputs. Each center has a radial unit that acts as the weight, whereas the radial distance of a feature value from the center is the threshold. The RBFN output drops rapidly toward zero for far-flung data since it observes a clumsy approach that limits the extrapolation of unknown data. Using a parzen window in this network makes it suitable for low-dimensional feature vectors. Faster learning, absence of error backpropagation mechanism, and local approximation make the classifier versatile for the recognition of BT images [Nanda *et al.*, 2023].

Figure 11.4 provides the basic RBFN structure. The network design is based on non-linear mapping of the input layer to the hidden layer from which it is linearly mapped to the output space. There are 'Z' patterns, each with a set of feature vectors, a single hidden layer with an H number of RBF units, and one output layer, as shown in Figure 11.4. The network provides an m-dimensional input feature space to one of the possible output spaces as given by

$$B : \mathbb{R}^m = \mathbb{R}^1 \tag{11.9}$$

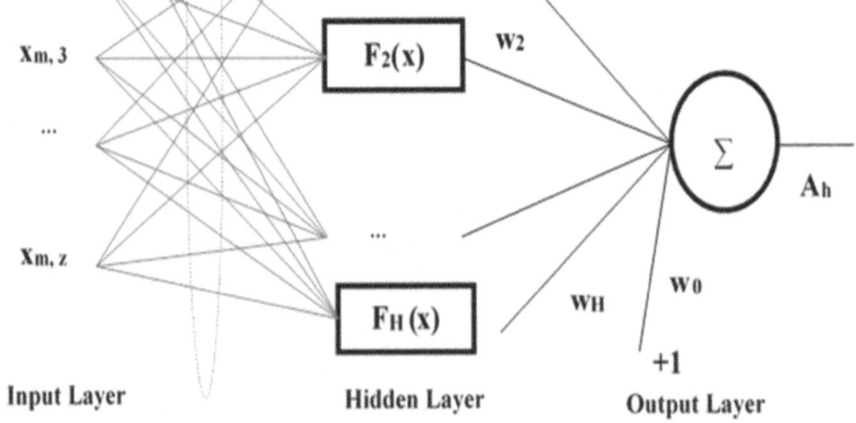

Figure 11.4 RBF Neural Network Structure.

where B denotes the chosen activation function. For a multi-class problem, several similar units are to be considered. The features from the chosen classes of BT images act as inputs to the network. The hidden units constitute the RBFs from these input feature vectors for classification tasks. Some distance measure '$\|\cdot\|$' need to be considered to choose the activation function $B(\cdot)$ as given by

$$B(\cdot) = \sum_{h=1}^{H} u_{m,h} F_h \|x - x_h\| \tag{11.10}$$

where H is the RBF index and $F_h \|x - x_h\|$ is the hidden RBF neurons, x_h is the RBF neuron centers, and $u_{m,h}$ is the weights corresponding to n^{th} input and the hth RBF unit. The Euclidean distance 'd' is similar to the RBF activation function, simpler, and is widely used in pattern recognition.

$$d = dist(x,h) = (x - x_h)^T \sum_{x}^{-1} (x - x_h) \tag{11.11}$$

The center x_h of the hth RBF functions can be computed corresponding to the extracted input feature streams through several techniques such as

- Select the extracted features randomly.
- Use some clustering algorithms such as self-organizing maps or K-means to fix the centroids.
- The default MATLAB orthogonal least squares algorithm has been used to select the centroids automatically.

This work considers the K-means technique to find the RBF centers. It segments the extracted input feature stream into different subsets while forming the desired clusters. The steps are as follows.

- Fix the number of RBFs and the number of clusters equal.
- Apply K-means clustering.
- Fix the number of RBFs centers and cluster centers equal.
- Fix the cluster variance and RBFS width equal.

Among several activation functions, the Gaussian function is localized and has better control and interpolation properties over the active region; hence, it is preferred for pattern classification [Sivanandan et al., 2007]. The network uses the Gaussian activation function $B(x)$ for h-RBF units, each with a width σ_h as given by

$$B(x_m) = \exp\left(-\frac{\|x - x_h\|^2}{2\sigma_h^2} \right) \tag{11.12}$$

The width σ signifies the input data distribution and indicates the smoothness of the activation function. The network output is a weighted sum of the RBF output units and is given by

$$A(x) = \sum_{h=1}^{H} B(x) w_h + w_0 \tag{11.13}$$

11.5 RESULTS AND DISCUSSION

A comparison of the RBFN performances between the WT and optimized feature vector has been provided in Table 11.2. Similarly, Table 11.3 compares the RBFN accuracy corresponding to the WT and optimized feature vectors. The optimized WT feature vectors using i-vectors outperformed the WT-based features due to more discriminating features, as observed in this table. In this work, several UBM Gaussian coefficients such as 16, 32, 64, 128, and 256 have been trailed to fetch the i-vectors. A value of 128 was chosen finally, as the effect of GMM counts was meagre after this value. The feature selection through the i-vector algorithm has reduced redundant data by mapping the high-dimensional raw WT feature vector into a reduced space, providing only relevant information, hence experiencing enhanced performances. Further, the i-vector aims to reduce the variability and normalize the covariance of the Gaussian super-vectors, thus providing

Table 11.2 Comparison of BT performance with proposed technique

Feature	*FN*	*TN*	*FP*	*TP*	*Specificity*	*Sensitivity*
WT	1045	6000	465	6650	92.80	86.42
Optimized WT	480	6200	153	7735	97.14	92.27

TP: True positive, TN: True Negative, FP: False positive, FN: False negative

Table 11.3 Comparison of BT Classification Accuracy with Proposed Technique

Feature Extraction	*Parameters*	*Average Training (%)*	*Average Testing (%)*
WT	Specificity	93.29	92.80
	Sensitivity	87.15	86.42
	Accuracy	89.72	89.3
Optimized WT	Specificity	98.02	97.14
	Sensitivity	93.89	92.27
	Accuracy	96.75	95.98

the desired discriminating features in classifying the chosen classes. The performance measures considered for the intended task are briefed below.

$$\text{Accuracy} = \frac{\text{TN} + \text{TP}}{\text{TP} + \text{FP} + \text{TN} + \text{FN}} \tag{11.14}$$

$$\text{Sensitivity} = \frac{\text{TP}}{\text{TP} + \text{FP}} \tag{11.15}$$

$$\text{Specificity} = \frac{\text{TN}}{\text{TN} + \text{FP}} \tag{11.16}$$

Table 11.4 compares the average training and testing accuracy versus number of epochs by simulating the RBFNN using the WT and optimized WT feature vectors simultaneously to understand the concept better.

Figure 11.5 provides the RBFNN loss graph for the proposed optimized feature vector with different number of epochs.

Table 11.4 A comparison of average training and testing accuracy of RBFNN between the WT and the optimized WT feature vectors

Average Training Accuracy					
Feature Vector	Number of Epochs				
	10	20	30	40	50
WT	73.45%	81.11%	84.53%	86.27%	89.72%
Optimized WT	80.22%	89.75%	93.45%	94.23%	96.75%
	Average Testing Accuracy				
	10	20	30	40	50
WT	67.19%	75.50%	80.66%	84.90%	89.3
Optimized WT	77.31%	85.02%	89.61%	92.88%	95.98%

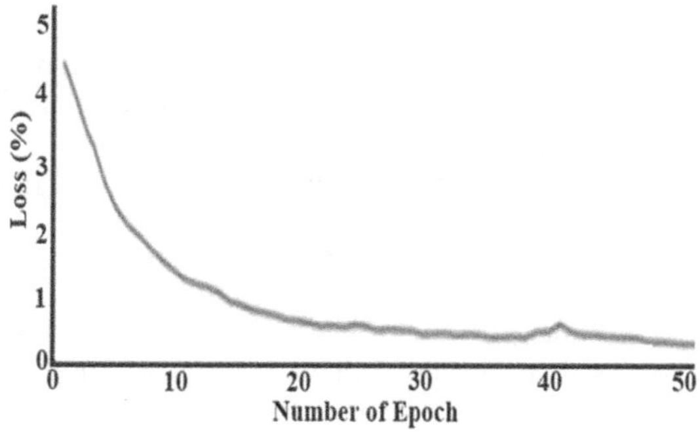

Figure 11.5 RBFNN loss graph.

Table 11.5 Comparison of computation time in seconds

Features Extraction	*Feature Extraction Time*	*Classification Time*	*Total Time*
Wavelet Transform	0.57s	0.44s	1.01s
Wavelet Transform + i-vector	0.89s	0.32s	1.21s

Table 11.5 compares the computation time at the feature extraction and classification stage at the illustrated approach. Observation shows that the WT takes less time to extract the feature than the proposed optimized WT due to the additional complexity of adapting the i-vector algorithm. Nevertheless, the classification time has been reduced as a tradeoff since the classifier requires training and validating the optimized feature vectors, which is low dimensional.

11.6 CONCLUSION

The central thrust area of this work is to extract reliable features of MR BT images through optimization that can lay new research directions. Using the i-vector feature reduction algorithm with WT has indeed increased the classification accuracy compared to the baseline wavelet features, as observed from our results. However, it has increased the feature extraction time due to the complexity involved. Nevertheless, the reduced dimension of the proposed feature set has increased the classifier's speed of response as a tradeoff. The comparison of the proposed method with other state-of-the-art feature reduction algorithms can provide essential inputs in this field. Similarly, feature extraction algorithms through modification and combination platforms through different optimization algorithms may further enhance the classification performance and will be used in future work.

REFERENCES

Alanazi, M. F. et al. (2022). Brain tumor/mass classification framework using magnetic-resonance-imaging-based isolated and developed transfer deep-learning model. *Sensors*, 22(1), 372.

Arif, M., Ajesh, F., Shamsudheen, S., Geman, O., Izdrui, D., & Vicoveanu, D. (2022). [Retracted] Brain Tumor Detection and Classification by MRI Using Biologically Inspired Orthogonal Wavelet Transform and Deep Learning Techniques. *Journal of Healthcare Engineering*, 2022(1), 2693621.

Bahador, N., & Kortelainen, J. (2022). Deep learning-based classification of multichannel bio-signals using directedness transfer learning. *Biomedical Signal Processing and Control*, 72, 103300.

Bishop, C. M. (1994). Neural networks and their applications. *Review of Scientific Instruments*, 65(6), 1803–1832.

Cheng, J., Huang, W., Cao, S., Yang, R., Yang, W., Yun, Z., ... & Feng, Q. (2015). Enhanced performance of brain tumor classification via tumor region augmentation and partition. *PLoS One*, 10(10), e0140381.

Cheng, J., Yang, W., Huang, M., Huang, W., Jiang, J., Zhou, Y., ... & Chen, W. (2016). Retrieval of brain tumors by adaptive spatial pooling and fisher vector representation. *PLoS One*, 11(6), e0157112.

Devi, R. S., Perumal, B., & Rajasekaran, M. P. (2022). A hybrid deep learning based brain tumor classification and segmentation by stationary wavelet packet transform and adaptive kernel fuzzy c means clustering. *Advances in Engineering Software*, 170, 103146.

El-Dahshan, E. S. A., Mohsen, H. M., Revett, K., Salem, A. B. M.: Computer-aided diagnosis of human brain tumor through MRI: A survey and a new algorithm. *Expert systems with Applications*, 41(11), 5526–5545 (2014).

Giraddi, S., Vaishnavi, S. V.: Detection of Brain Tumor using Image Classification. In *2017 International Conference on Current Trends in Computer, Electrical, Electronics and Communication (CTCEEC)* (pp. 640–644), IEEE (2017, September).

Gupta, D., & Ahmad, M. (2018). Brain MR image denoising based on wavelet transform. *International Journal of Advanced Technology and Engineering Exploration*, 5(38), 11–16.

Iftekharuddin, K. M., & Parra, C. (2003, November). Multiresolution-fractal feature extraction and tumor detection: analytical model and implementation. In *Wavelets: Applications in Signal and Image Processing X* (Vol. 5207, pp. 801–812). SPIE.

Kaur, T., Saini, B. S., & Gupta, S. (2018). A novel feature selection method for brain tumor MR image classification based on the Fisher criterion and parameter-free Bat optimization. *Neural Computing & Applications*, 29(8), 193–206.

Kaya, I. E., Pehlivanlı, A. Ç., Sekizkardeş, E. G., & Ibrikci, T. (2017). PCA based clustering for brain tumor segmentation of T1w MRI images. *Computer Methods and Programs in Biomedicine*, 140, 19–28.

Khandan, F., & Ayatollahi, A. (2003, September). Performance region of center affine filter for eliminating of interference terms of discrete Wigner distribution. In *3rd International Symposium on Image and Signal Processing and Analysis, 2003. ISPA 2003. Proceedings of the (Vol. 2, pp. 621-625)*. IEEE.

Kumar, S., Dabas, C., & Godara, S. (2017). Classification of brain MRI tumor images: a hybrid approach. *Procedia Computer Science*, 122, 510–517.

Lakshmipriya, B., & Jayalakshmy, S. (2023). Wavelet scattering and scalogram visualization based human brain decoding using empirical wavelet transform. *International Journal of Information Technology*, 15(3), 1699–1708.

Majid, M. A., Abidin, A. F. Z., Anuar, N. D. K., Kadiran, K. A., Karis, M. S., Yusoff, Z. M., ... & Rizman, Z. I. (2018). A comparative study on the application of binary particle swarm optimization and binary gravitational search algorithm in feature selection for automatic classification of brain tumor MRI. *Journal of Fundamental and Applied Sciences*, 10(2S), 486–498.

Mastrogiacomo, S., Dou, W., Jansen, J. A., & Walboomers, X. F. (2019). Magnetic resonance imaging of hard tissues and hard tissue engineered bio-substitutes. *Molecular Imaging and Biology*, 21(6), 1003–1019.

Mohanty, M. N., & Palo, H. K. (2020). Child emotion recognition using probabilistic neural network with effective features. *Measurement*, 152, 107369.

Mohanty, M. N., & Palo, H. K. (2019). Segment based emotion recognition using combined reduced features. *International Journal of Speech Technology*, 22(4), 865–884.

Nanda, A., Barik, R. C., & Bakshi, S. (2023). SSO-RBNN driven brain tumor classification with Saliency-K-means segmentation technique. *Biomedical Signal Processing and Control*, 81, 104356.

Nawaz, S. A., Khan, D. M., & Qadri, S. (2022). Brain tumor classification based on hybrid optimized multi-features analysis using magnetic resonance imaging dataset. *Applied Artificial Intelligence*, 36(1), 2031824.

Oppenheim, A. V., Schafer, R. W., Rabiner, L. R., Gold, B., & Hunt, B. R. (1976). Digital signal processing and theory and application of digital signal processing, *Physics Today*, 29(1), 63–64.

Palo, H. K., Subudhiray, S., & Das, N. (2023). The amalgamation of wavelet packet information gain entropy tuned source and system parameters for improved speech emotion recognition. *Speech Communication*, 149, 11–28.

Palo, H. K., Mohanty, M. N., & Chandra, M. (2016). Efficient feature combination techniques for emotional speech classification. *International Journal of Speech Technology*, 19, 135–150.

Palo, H. K., & Sarangi, L. (2020). Overview of machine learners in classifying of speech signals. In *Handbook of research on emerging trends and applications of machine learning* (pp. 461–489). IGI Global.

Rai, H. M., & Chatterjee, K. (2019). Hybrid adaptive algorithm based on wavelet transform and independent component analysis for denoising of MRI images. *Measurement*, 144, 72–82.

Rathi, V. P., & Palani, S. (2012). Brain tumor MRI image classification with feature selection and extraction using linear discriminant analysis. arXiv preprint arXiv:1208.2128.

Sahoo, L., Sarangi, L., Dash, B. R., Palo, H. K. (2020). Detection and Classification of Brain Tumor Using Magnetic Resonance Images. In *Advances in Electrical Control and Signal Systems: Select Proceedings of AECSS 2019* (pp. 429–441). Springer Singapore.

Samal, L., Palo, H. K., & Sahu, B. N. (2023). The recognition of 3-phase power quality events using optimal feature selection and random forest classifier. *International Journal of Computational Vision and Robotics*, 13(3), 235–246.

Samal, L., Palo, H. K., & Sahu, B. N. (2020, July). Comparison of classifiers for power quality disturbances with wavelet statistical analysis. In *2020 International Conference on Computational Intelligence for Smart Power System and Sustainable Energy (CISPSSE)* (pp. 1–5). IEEE.

Sathish, P., & Elango, N. M. (2022). Gaussian hybrid fuzzy clustering and radial basis neural network for automatic brain tumor classification in MRI images. *Evolutionary Intelligence*, 15(2), 1359–1377.

Shree, N. V., & Kumar, T. N. R. (2018). Identification and classification of brain tumor MRI images with feature extraction using DWT and probabilistic neural network. *Brain Informatics*, 5(1), 23–30.

Sharma, Y., & Singh, B. K. (2023). Classification of Children With Attention-Deficit Hyperactivity Disorder Using Wigner-Ville Time-Frequency and Deep expEEGNetwork Feature-Based Computational Models. *IEEE Transactions on Medical Robotics and Bionics*, 5(4), 890–902.

Singh, A.: Detection of brain tumor in MRI images, using combination of fuzzy c-means and SVM. In *2015 2nd International Conference on Signal Processing and Integrated Networks (SPIN)*, (pp. 98–102), IEEE (2015, February).

Sivanandam, S. N., & Deepa, S. N. (2007). *Principles of soft computing (with CD)*. John Wiley & Sons.

Subudhiray, S., Palo, H. K., & Das, N. (2023). Effective recognition of facial emotions using dual transfer learned feature vectors and support vector machine. *International Journal of Information Technology*, 15(1), 301–313.

Taebi, A., & Mansy, H. A. (2017, December). Analysis of seismocardiographic signals using polynomial chirplet transform and smoothed pseudo Wigner-Ville distribution. In *2017 IEEE Signal Processing in Medicine and Biology Symposium (SPMB)* (pp. 1–6). IEEE.

Tang, B., Liu, W., & Song, T. (2010). Wind turbine fault diagnosis based on Morlet wavelet transformation and Wigner-Ville distribution. *Renewable Energy*, 35(12), 2862–2866.

Tahosin, M. S., Sheakh, M. A., Islam, T., Lima, R. J., & Begum, M. (2023). Optimizing brain tumor classification through feature selection and hyperparameter tuning in machine learning models. *Informatics in Medicine Unlocked*, 43, 101414.

Chapter 12

Enhancing breast cancer detection: A deep transfer learning approach with AlexNet

Rajen Bose, Bidushkumar Sahoo, and Apul N. Dev

12.1 INTRODUCTION

A complex nature of breast cancer is demonstrated by the genetic, hormonal, environmental, and lifestyle risk factors [Hassan & El-Shenawee, 2011]. Increasing survival rates and enabling prompt intervention depend on early detection by screening mammography, clinical breast examinations, and self-examinations. Accuracy and efficiency of breast cancer diagnosis have been claimed to be improved by combining DL algorithms with medical imaging in recent years. The ability of DL to apply previously acquired information from enormous datasets to novel fields has drawn a lot of attention [Ganesan et al. (2012), Sun et al. (2017), Khuriwal & Mishra (2018)]. This article presents a revolutionary approach to breast cancer diagnostics that integrates deep learning with the foundational convolutional neural network architecture, Alex Net. We intend to increase the sensitivity and specificity of current screening methods by using Alex Net's exceptional feature extraction capabilities and optimizing its parameters on a big collection of mammographic pictures [Nawaz et al., 2018]. Several benefits come with using Alex Net in transfer learning for medical image processing. First of all, the multiple convolutional and pooling layers in AlexNet's architecture enable the extraction of fine patterns and features that are essential for differentiating between benign and malignant breast cancers. Second, a crucial starting point, its pretrained weights were acquired from a sizable dataset of natural images, enable quick convergence and do away with the requirement for a significant amount of training data in the medical field. Using transfer learning, which entails retraining the network on a specific dataset of mammograms, we modify AlexNet to the task of breast cancer detection. The model is confirmed to be robust and generalizable across a variety of clinical situations by the large range of breast tissue compositions, lesion types, and imaging modalities included in this dataset.

 DOI: 10.1201/9781003589716-12

12.1.1 BC detection using ML

Breast cancer, one of the most common malignancies in women, highlights the importance of early detection and precise diagnosis in enhancing survival and treatment results. Mammography, biopsy, and physical examination are still frequently employed, but delays, interpretation variability, and invasive procedures limit them. By using computational algorithms to evaluate complicated data patterns, machine learning (ML) is revolutionizing breast cancer diagnosis. ML can automate and optimize diagnostic processes for earlier diagnosis, better accuracy, and individualized therapy. ML systems understand patterns and anticipate new data by training on large datasets of malignant and benign cases. Breast cancer detection is especially effective using several ML methods. SVMs, decision trees, random forests, and neural networks are popular supervised learning approaches. These algorithms distinguish malignant versus non-cancerous cases using labeled training data like patient imaging or biopsy results. Random forests and SVMs excel at handling high-dimensional data and complex feature relationships. Convolutional neural networks (CNNs) excel in image analysis. CNNs automatically extract hierarchical information from mammograms, ultrasounds, and MRI scans to detect and classify tumors accurately.

Feature engineering is crucial to breast cancer detection with ML. Model accuracy depends on identifying and using key parameters including tumor size, shape, density, texture, and hormone receptor status. PCA and RFE minimize data dimensionality and choose the most informative features. Ensemble approaches like bagging and boosting increase accuracy and prevent overfitting by combining model predictions. For breast cancer categorization, XGBoost and AdaBoost work well. Public datasets have sped ML model development and evaluation. Research has advanced thanks to popular datasets like WBCD and digital mammography datasets like MIAS and DDSM. These datasets contain labeled data for robust ML model training. Python, TensorFlow, Keras, and scikit-learn provide libraries and frameworks for implementing, training, and testing ML algorithms. ML integration into clinical practice is difficult despite its potential. Privacy considerations make high-quality data scarce, and institution-wide data harmonization is difficult. Many ML models, especially deep learning ones, are black boxes, making it hard for physicians to comprehend and use their predictions. Small or imbalanced datasets can also cause overfitting, when models perform well in training but poorly in practice.

Continuous efforts are made to address these issues. Clinicians may trust and utilize this technology because researchers are creating interpretable AI models that provide clear, actionable insights into their decision-making processes. Integrating imaging, genetic, and clinical data for a more complete patient picture is another intriguing approach. Federated learning is being investigated for collaborative model training while protecting patient privacy. Machine learning could revolutionize breast cancer detection by improving diagnosis, reducing errors, and personalizing treatment. ML can improve healthcare outcomes, reduce worldwide breast cancer mortality, and improve millions of patients' quality of life by overcoming current barriers and encouraging interdisciplinary collaboration between data scientists, physicians, and researchers.

12.1.2 BC detection using DL

Early and precise detection is essential for efficient treatment and better survival rates for breast cancer, the leading cause of cancer deaths in women worldwide. Mammography, ultrasonography, and biopsy are useful but time-consuming and prone to human mistake. Deep learning is a groundbreaking healthcare technique that detects breast cancer with unmatched precision and speed. Deep learning can evaluate complicated medical data patterns using neural networks that imitate human brain function to detect and diagnose early. CNNs lead breast cancer diagnosis. CNNs have been frequently utilized to analyze mammograms, ultrasounds, and MRIs because they process grid-like data structures. Without manual feature engineering, these networks extract hierarchical characteristics like edges, forms, textures, and patterns from incoming data. CNNs can detect microcalcifications and lumps in mammograms, which indicate early-stage breast cancer. CNNs are effective at breast cancer categorization, tumor segmentation, and lesion detection because they can learn complex representations.

DL can handle big, high-dimensional datasets, making it useful for breast cancer detection. The Digital Database for Screening Mammography (DDSM), Breast Cancer Digital Repository (BCDR), and INbreast have helped design and train robust deep learning models. These datasets include annotated photos to help models accurately discriminate malignant and benign instances. The use of pretrained models like VGGNet, ResNet, and Inception reduces the computational cost and time needed to train models for breast cancer detection applications because to transfer learning breakthroughs. Integration of multimodal data is also possible with deep learning. Models can examine histopathology slides, genetic data, and EHRs in addition to imaging to provide comprehensive patient insights. Combining imaging data with histopathology analysis using deep learning can increase diagnostic accuracy and eliminate false positives and negatives.

DL in breast cancer diagnosis is difficult despite its potential. Training requires huge, annotated datasets, a major difficulty. Medical image annotation is laborious and needs radiologists, restricting labeled data. Deep learning models are also black boxes, making their decision-making process difficult for clinicians to understand. Their lack of openness can limit clinical adoption. To understand model decision-making, researchers are employing explainable AI (XAI) methodologies. Saliency maps and Grad-CAM emphasize medical image regions that influence model prediction, enhancing trust and usability.

Model generalizability is another issue. When applied to data from diverse populations or imaging devices, deep learning models trained on one demographic or device may perform badly. Domain adaptation and data augmentation improve model resilience and consistency across varied datasets.

Federated learning frameworks that allow models to be trained on decentralized data from many institutions without compromising patient privacy are future deep learning directions for breast cancer detection. Radiologists can receive rapid diagnostic support from real-time deep learning algorithms, decreasing workloads and improving decision-making.

DL is changing breast cancer diagnosis by evaluating medical pictures and multimodal data with high accuracy and efficiency. Deep learning could enhance breast cancer detection, patient outcomes, and worldwide mortality by addressing data restrictions, interpretability, and generalization.

12.1.3 BC detection using XAI

The global health burden of breast cancer requires early identification to improve patient outcomes and reduce mortality. Diagnostic methods including mammography, ultrasound, and biopsies require radiologists and pathologists, which can be time-consuming and error-prone. AI, especially deep learning, has improved breast cancer diagnosis accuracy and efficiency. However, many AI models are black boxes, making forecasts without revealing decision-making processes. Lack of interpretability can harm clinician trust and adoption. Explainable AI (XAI) could solve this problem by making AI predictions more clear, interpretable, and clinically relevant. Saliency maps, Grad-CAM, SHapley Additive Explanations, and LIME reveal how AI models make decisions. Saliency maps and Grad-CAM show crucial spots in medical images like mammograms where the model predicts anomalies, allowing radiologists to evaluate results and discover microcalcifications or tumors. SHAP and LIME measure how tumor size, shape, and texture affect the model's predictions, helping doctors comprehend diagnostic outcomes. These methods align AI-assisted diagnostics with clinical experience and enable collaborative decision-making to build trust. Beyond trust, XAI improves diagnostic accuracy, mistake reduction, tailored treatment planning, and ethical and regulatory compliance. XAI allows individualized care techniques like tumor-based therapy aggressiveness by detecting and explaining crucial predictors. XAI also helps detect biases or flaws in AI models, improving refinement and generalizability across varied patient groups. XAI for breast cancer diagnosis is difficult, despite its potential. Radiologists must grasp clinically appropriate interpretations of complicated deep learning models, which can be computationally difficult. Oversimplified explanations may lose important subtleties, making it difficult to reconcile accuracy and interpretability. Future XAI advances will integrate these technologies into real-time diagnostic workflows to help radiologists make quick choices. Federated learning techniques, which train explainable models across decentralized datasets while protecting patient privacy, can scale healthcare XAI systems. XAI addresses the requirement for transparency and improves AI-driven breast cancer detection systems by merging interpretability with cutting-edge AI. This combination of technology and clinical skills could transform cancer diagnosis, patient care, and breast cancer burden worldwide.

The potential of our work to address the shortcomings of traditional breast cancer screening techniques—limited sensitivity, high false-positive rates, and subjective interpretation—makes it significant. Our goal is to increase the precision and effectiveness of the diagnosis of breast cancer by integrating DL with AlexNet, therefore enabling earlier identification and treatment. We provide in this work a thorough explanation of our DL method, experimental setting, and evaluation results for breast cancer diagnosis using AlexNet. Moreover, we stress the significance of our results and directions for further investigation, stressing the revolutionary potential of artificial intelligence in improving the treatment of breast cancer.

12.2 LITERATURE REVIEW

Many research has looked at using DL methods especially CNNs to interpret medical imaging data, including mammograms and breast ultrasound scans. There has been a lot of study done to assess how well different ML and DL algorithms work for breast cancer detection.

Rathi and Pareek (2016) put forth a hybrid model that incorporates four classifiers (SVM, Naïve Bayes, Decision tree) with MR feature selection. The most effective classifier, they found, was SVM. As noted by Bharat et al., KNN had higher accuracy than SVM, decision trees, and Naïve Bayes classifiers. On the Wisconsin breast cancer dataset, Bayrak et al. (2019) evaluated many ML methods. The researchers found that in terms of performance measures, SVM proved to be superior to other classifiers. Shravya et al. (2019) switched from ML to DL. They concentrated on methods of supervised ML like KNN, SVM, and Logistic Regression, and especially, SVM performed its analysis with remarkable accuracy. In their comparison of Random Forest, SVM, Logistic Regression, and Naïve Bayes, Sivapriya et al. (2019) found that Random Forest outperformed all other models in terms of accuracy. Moreover, Sornam et al. (2017) emphasized the applicability of DL methods for picture identification in a survey including several fields. The performance of classifiers with and without feature selection techniques was compared by Deepa et al. in their presentation of a method for enhancing them. On the Wisconsin breast cancer dataset, Soni et al. (2019) obtained 99.714% accuracy using SVM and random forest techniques. Better understanding of ML and DL approaches for breast cancer detection is made possible by this work. For the greatest outcomes, they stress the significance of carefully choosing algorithms and technical components. Ghosh et al. (2014) put up an independent component analysis method for reducing dimensionality. Using ANN, KNN, RBFNN, and SVM among other ML techniques, they then applied this method to the Wisconsin Diagnostic Breast Cancer dataset. ANN, KNN, SVM, and RBFNN that used this method obtained accuracy of 97.53%, 91.03%, 95.25%, and 90.49%. Zheng et al. (2014) thoroughly examined six techniques using the WDBC dataset. Additionally, to Linear Regression, Multilayer Perceptron, Nearest Neighbor Search, Softmax Regression, and SVM," this study used a Gated Recurrent Unit and SVM ensemble model. The MLP obtained the highest accuracy of 99.04% out of all the algorithms examined. These results clarify various ML techniques and dimensionality reduction techniques in the context of breast cancer detection. They emphasize the potential of higher accuracy with the application of appropriate algorithms.

Gene signatures were tried to be used by Jha et al. (2018) to forecast breast cancer relapse. The work used the special properties of genes by using a Graph CNN model. The findings indicated that GCNN was more accurate in predicting relapse than earlier algorithms. Srirambabu et al. (2010) have presented a model for breast cancer diagnosis based on the SVM classifier. Several procedures were taken in turn, such as feature extraction, segmentation, and picture improvement, before an SVM classifier was used. The data used in this work was the MIAS database. Segmentation was achieved with thresholding methods and noise reduction via median filtering. Medjahed et al. (2013) presented a KNN algorithm-based breast cancer classification model. With the Wisconsin Breast Cancer dataset, the authors used distance measures to evaluate the model's performance. Accuracy values for Euclidean distance were 98.70%, while for Manhattan distance they were 98.48%. These research shows how well ML and neural network approaches use genetic and imaging data to improve diagnostic precision, so advancing breast cancer detection methods.

There are several drawbacks to take into account even although the studies mentioned above described several effective methods for DL-based breast cancer detection. These include difficulties with interpretability and explainability of the DL models, which are important in clinical decision-making, as well as possible biases in the datasets utilized and restricted generalizability to different patient populations or imaging modalities.

12.2.1 Advantages and limitations of ML and DL methods for BC detection

Table 12.1 shows the merits and weaknesses of ML and DL techniques to BC screening, highlighting areas for development.

12.3 MATERIALS AND METHODS

The AlexNet model, a deep CNN, improves breast cancer detection through picture categorization, decreasing human interpretation and paving the door

Table 12.1 Advantages and limitations of ML and DL methods in BC detection

Author(s)	*Methodology*	*Advantages*	*Limitations*
Rathi and Pareek (2016)	Hybrid model with SVM, Naïve Bayes, Decision Tree, and MRMR feature selection	Enhanced accuracy with MRMR feature selection; SVM identified as the most effective classifier.	Computationally expensive; reliance on feature selection for optimal performance.
Bharat et al. (2018)	KNN compared with SVM, Decision Trees, and Naïve Bayes	KNN outperformed other models in accuracy; straightforward implementation.	Limited scalability for high-dimensional datasets.
Bayrak et al. (2019)	Evaluation of ML methods using the Wisconsin Breast Cancer Dataset	SVM demonstrated superior performance; robust on structured datasets.	Limited generalizability to other datasets; dataset-specific results.
Shravya et al. (2019)	Supervised ML methods (KNN, SVM, Logistic Regression)	High accuracy with SVM; transition to DL approaches for better feature learning.	Focused on ML models; lacks exploration of DL models.
Sivapriya et al. (2019)	Random Forest compared with SVM, Logistic Regression, and Naïve Bayes	Random Forest achieved the highest accuracy; robust against overfitting.	Requires large datasets for effective training; interpretability challenges.
Sornam et al. (2017)	Survey on DL methods for image-based detection	Highlighted the applicability of DL for diverse tasks; strong capability for image recognition.	Generalized insights; lacks specific experimental validation.
Deepa et al. (2019)	Comparison of classifiers with and without feature selection	Demonstrated enhanced accuracy with feature selection; presented techniques for optimization.	Heavy reliance on feature selection; performance variability across datasets.
Soni et al. (2019)	SVM and Random Forest on the Wisconsin Breast Cancer Dataset	Achieved 99.714% accuracy; combined use of SVM and Random Forest improved classification.	Limited to one dataset; overfitting potential on small datasets.

(*Continued*)

Table 12.1 (Continued)

Author(s)	*Methodology*	*Advantages*	*Limitations*
Ghosh et al. (2014)	Independent Component Analysis (ICA) for dimensionality reduction with ML techniques (ANN, KNN, SVM, RBFNN)	Enhanced accuracy with ICA; ANN and SVM outperformed other methods.	Complexity of ICA; model dependence on dimensionality reduction quality.
Zheng et al. (2014)	Gated Recurrent Unit and SVM ensemble; analysis of six techniques using the WDBC dataset	MLP achieved 99.04% accuracy; ensemble methods improved robustness.	Computationally intensive; complex ensemble approach.
Jha et al. (2018)	Graph CNN (GCNN) for predicting breast cancer relapse	High accuracy in relapse prediction; innovative use of gene signatures.	Limited to gene-specific datasets; challenges in scalability to broader applications.
Srirambabu et al. (2010)	SVM-based model with preprocessing steps like segmentation and noise reduction using MIAS database	Effective segmentation with thresholding; median filtering enhanced image quality.	Preprocessing is computationally heavy; depends on image quality and dataset diversity.
Medjahed et al. (2013)	KNN with distance measures on the Wisconsin Breast Cancer Dataset	High accuracy with Euclidean and Manhattan distance measures; simplicity in implementation.	Sensitive to noise and outliers; less effective for high-dimensional data.

for automated diagnostic tools. AlexNet, known for its capacity to process complicated visual data, can evaluate subtle patterns and textures, making it perfect for breast cancer detection. Hierarchical feature extraction allows the model accurately distinguish between normal and malignant tumors.

The system starts with raw picture capture from the broad and high-quality IRMA dataset. Raw photos are preprocessed to meet AlexNet's input criteria. Steps involve scaling photos to a fixed resolution (e.g., 227 × 227) and converting them to RGB channels to meet AlexNet model input parameters. Preprocessing can also use image augmentation methods including rotation, flipping, and contrast modifications to increase dataset variety and decrease overfitting. This phase ensures the model is robust and works effectively on unknown data. AlexNet classifies input photos using its convolutional, pooling, and fully connected layers after preprocessing. Convolutional layers collect edges and textures, whereas deeper layers capture high-level patterns and structures to distinguish normal and cancerous breast tissues. The final output layer assigns probability scores to normal and malignant classes based on the input image.

Standard criteria including accuracy, precision, recall, and F1-score are used to evaluate AlexNet-based systems. These measures demonstrate the system's reliability and resilience in detecting malignant tumors and minimizing false positives. Analysts examine the model's confusion matrix to better understand classification errors and make targeted changes. In breast cancer detection, AlexNet has many benefits. Its pretrained architecture, frequently fine-tuned on huge medical datasets, speeds up training and improves accuracy. Transfer learning helps AlexNet classify breast cancer with little data by using expertise from existing datasets (Figure 12.1).

12.3.1 AlexNet

Three max-pooling layers, three fully connected (FC) layers, and five convolutional layers make up the innovative CNN AlexNet, whose architectural arrangement is shown in Figure 12.2. Following the first two convolutional layers, the first two max-pooling layers are applied in a sequential order; the other convolutional layers are connected directly. We introduce a third

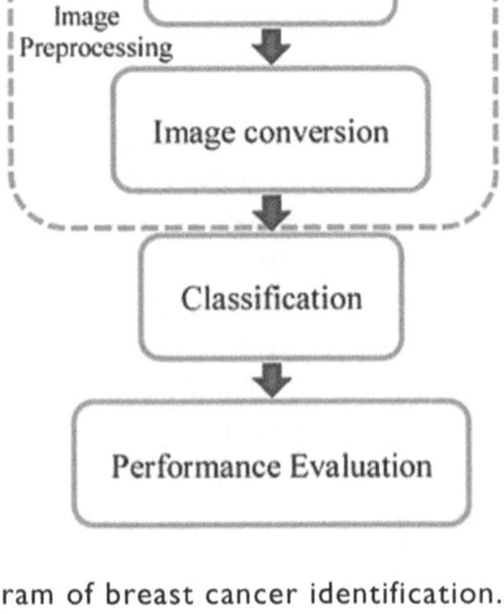

Figure 12.1 Block diagram of breast cancer identification.

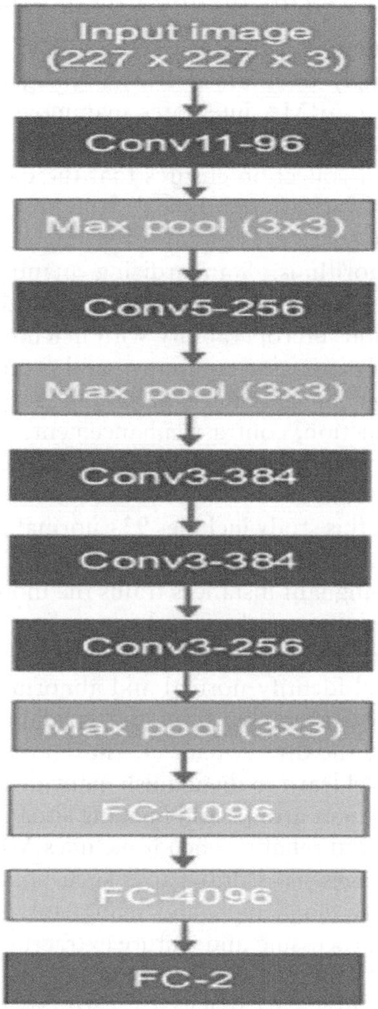

Figure 12.2 AlexNet architecture.

max-pooling layer after the fifth convolutional layer. Three fully linked layers receive this layer's output after that. The performance of AlexNet is frequently credited to significant advances in dropout regularization, "stochastic gradient descent (SGD), and Rectified Linear Units (ReLU)." Training is improved by the con-vergence of the model accelerated by the ReLU activation functions. The SGD method lowers the total cost function, hence optimizing the convolutional kernel parameters. To lessen overfitting, dropout regularization is applied in the first two completely connected layers. Known by another name, the softmax layer, the third FC layer in the CNN architecture categorizes objects [Han et al. (2017)].

12.3.2 Database description and image preprocessing

This study used mammography images from RWTH Aachen University, Germany's painstakingly maintained IRMA (Image Retrieval in Medical Applications) database. IRMA integrates mammogram data from RWTH Aachen routine images, MIAS, DDSM, and LLNL into a comprehensive repository. This varied collection ensures that the database includes breast tissue photographs of various anatomical structures, imaging settings, and clinical situations. Diversity is essential for strong and generalizable breast cancer detection algorithms. Standardizing mammography image segments from the IRMA database to a uniform resolution of 128x128 pixels ensures consistency and interoperability with machine learning algorithms. Standardization reduces imaging equipment and protocol variability, allowing models to focus on clinically relevant mammography patterns and characteristics. Noise reduction, contrast enhancement, and normalization can improve image quality and emphasize diagnostic characteristics such microcalcifications and masses.

The dataset used in this study includes 931 normal (benign) and 584 aberrant (malignant) pictures [Oliveira et al. (2008)]. This balanced representation of benign and malignant instances trains the model on enough samples from both groups, minimizing bias and improving generalization to new data. The image annotations give ground truth labels for supervised learning, helping the model identify normal and abnormal mammograms. This study uses the IRMA database to train and validate deep learning models like AlexNet on a vast and diverse dataset. The database's variety of imaging sources helps the model learn to distinguish many disease patterns, boosting its resilience across patient groups and imaging situations. The training data is also more credible and reliable when it includes MIAS and DDSM data.

Mammography images have drawbacks such noise, contrast levels, and small distinctions between benign and malignant instances. These issues require advanced preprocessing and feature extraction to improve diagnosis accuracy. The dataset's limited size (931 normal and 584 aberrant photos) may further limit the model's capacity to learn complicated patterns. This constraint can be overcome using data augmentation, transfer learning, and cross-validation to maintain model performance.

In conclusion, the IRMA database is ideal for developing advanced breast cancer detection algorithms. Its complete and standardized mammography pictures enable strong AI model training, improving diagnostic precision and possibly early identification and treatment planning. Future studies could include imaging modalities or use federated learning to access varied datasets while protecting patient privacy.

Image preprocessing consists of two essential stages: image conversion and image scaling. As the initial pixel size is insufficient to conform the images to the input layer dimensions of the network, resizing is performed. The images are resized to dimensions of 224 × 224 pixels in order to maintain consistent input image sizes for the classifier. Furthermore, image conversion is utilized to standardize the images, ensuring they have a three-channel input format that is necessary for the network to function properly. To achieve the necessary three-channel input configuration, grayscale images are transformed into RGB format. Moreover, the dataset is divided into training and validation subsets. 30% of the images are assigned to the training dataset, while the remaining 70% are allocated to the validation dataset. In order to minimize prejudice and guarantee impartial outcomes, the division is conducted in a random manner. Figure 12.2 displays representative samples from each class of the IRMA database. The samples of normal and abnormal cancer cases are displayed in the rows of Figure 12.3.

12.4 EXPERIMENTAL DESIGN AND RESULT ANALYSIS

PyCharm, a popular Python IDE, was used to construct the AlexNet breast cancer detection system. Python 3.6 was used for its library and framework compatibility. Windows 11 provided a robust environment for machine learning and deep learning research and testing. Libraries like Scipy, Numpy, and OpenCV were employed for efficient numerical computations and data handling. For efficient mammography image processing, these libraries provide a solid basis for image processing and multidimensional arrays. The device accelerates computing with an 8 GB RAM GPU. GPUs reduce

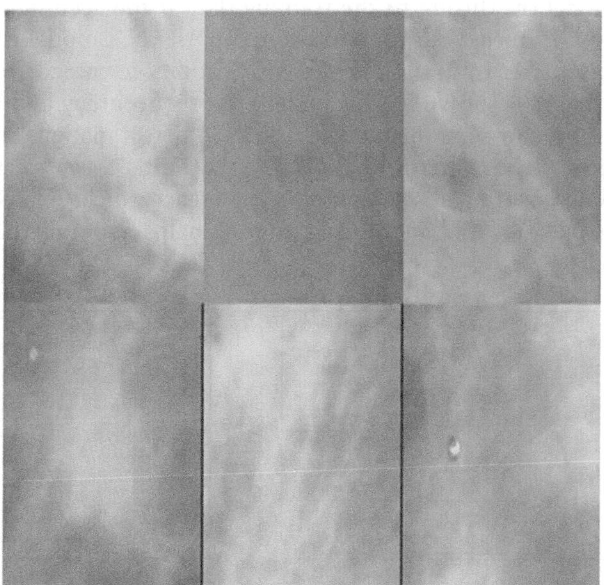

Figure 12.3 Sample of mammogram patches.

training time for deep learning models by performing matrix operations and parallel processing quicker than CPUs. This framework uses Keras, a TensorFlow-based high-level neural network API, for deep learning. Keras simplifies model-building and facilitates speedy experimentation, making it excellent for prototyping and testing.

The AlexNet architecture was used in the model design because of its picture categorization success. All layers of the network were optimized using ReLU activation functions. ReLU allows faster convergence during training by mitigating the vanishing gradient problem, making it popular for deep learning models. RMSProp (Root Mean Squared Propagation) optimized training on gradient-varying data by adapting the learning rate for each parameter. A learning rate of 0.0001 ensured consistent training, preventing significant updates from destabilizing the model. The network learned relevant features from mammography pictures after 100 epochs of training. Each epoch involved a forward network pass and a backward pass for weight updates via backpropagation. To assess model performance, accuracy and loss curves were monitored during training. The loss curve illustrates the model's mistake rate and improvement with time, while the accuracy curve shows how successfully it classifies images. These curves show the model's learning process and can indicate overfitting or underfitting.

Figure 12.4 shows accuracy and loss over 100 epochs of training. The network is learning from data when a well-behaved model steadily increases accuracy and decreases loss. Regularization or early halting may be needed to fix the model overfitting to the training data if the accuracy plateaus or loss increases. To measure generalization, the AlexNet model is tested on a distinct dataset after training. The network's ability to categorize new mammography images is shown by the model's test set accuracy and loss. Testing is essential for confirming the model's ability to detect malignant and benign tumors in real life. PyCharm, Python 3.6, the Keras framework, and optimized training ensure a stable and efficient breast cancer detection system. Monitoring accuracy and loss during training helps fine-tune the model and

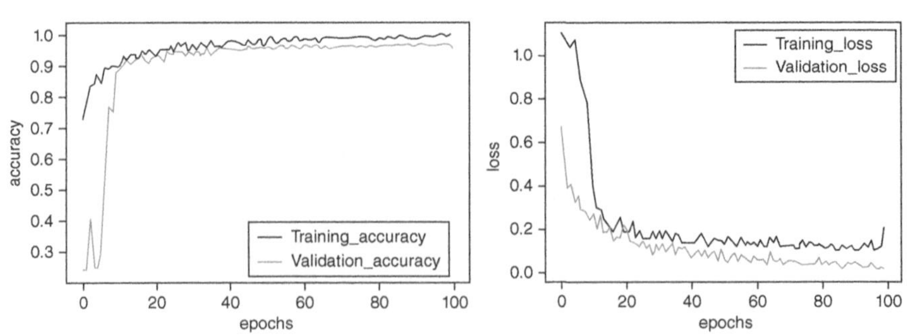

Figure 12.4 Accuracy and loss graph for AlexNet.

Class Name	Precision	1-Precision	Recall	1-Recall	f1-score
Class0	0.9857	0.0143	0.9892	0.0108	0.9875
Class1	0.9830	0.0170	0.9774	0.0226	0.9802
Accuracy	0.9846				
Misclassification Rate	0.0154				
Macro-F1	0.9838				
Weighted-F1	0.9846				

Figure 12.5 Performance measures for AlexNet.

improve its performance, making it a more trustworthy diagnostic tool. To improve model resilience and accuracy, future work may involve experimenting with other architectures, tweaking hyperparameters, or adding more advanced data augmentation approaches.

The recall, precision, F1-score, sensitivity, and specificity metrics were used to evaluate the performance of the AlexNet model; the results are shown in Figure 12.5. Figure 12.6 shows the AlexNet confusion matrix. The following formulas were used to calculate the performance metrics.

12.5 CONCLUSION

DL applied with the AlexNet architecture has shown to have great promise for improving the efficiency and accuracy of breast cancer detection. The objective of our work was to improve the sensitivity and specificity of current

Training Set			
TARGET / OUTPUT	Class0	Class1	SUM
Class0	276 60.53%	4 0.88%	280 98.57% 1.43%
Class1	3 0.66%	173 37.94%	176 98.30% 1.70%
SUM	279 98.92% 1.08%	177 97.74% 2.26%	449 / 456 98.46% 1.54%

Figure 12.6 AlexNet confusion matrix.

screening techniques by using the strong feature extraction capabilities of AlexNet and optimizing its settings on a large dataset of mammographic pictures. By use of preprocessing methods and AlexNet classification, we were able to classify mammography pictures into normal or aberrant tumor categories. A high accuracy grade of 98.46% obtained from the evaluation of our method demonstrated the usefulness of DL in medical picture analysis. Furthermore, confirming the robustness of the AlexNet model in precisely detecting breast cancer cases was the performance metrics recall, precision, F1-score, sensitivity, and specificity. Our results have important ramifications for raising early identification and treatment of breast cancer, which will eventually lead to better patient outcomes and survival rates. Going forward, more study and development in this field might result in the clinical practice integrating DL technology, completely changing the way breast cancer is treated.

REFERENCES

Bayrak, E. A., Kırcı, P., &Ensari, T. (2019, April). Comparison of ML methods for breast cancer diagnosis. In *2019 Scientific meeting on electrical-electronics & biomedical engineering and computer science (EBBT)* (pp. 1–3). IEEE.

Bharat, A., Pooja, N., & Reddy, R. A. (2018, October). Using ML algorithms for breast cancer risk prediction and diagnosis. In *2018 3rd International conference on circuits, control, communication and computing (I4C)* (pp. 1–4). IEEE.

Deepa, B. G., Senthil, S., Gupta Rahil, M., & Shah Vishakha, R. (2019). Augmentation of classifier accuracy through implication of feature selection for breast cancer prediction. *International Journal of Recent Technology and Engineering*, 8(2), 6396–6399.

Ganesan, K., Acharya, U. R., Chua, C. K., Min, L. C., Abraham, K. T., & Ng, K. H. (2012). Computer-aided breast cancer detection using mammograms: a review. *IEEE Reviews in Biomedical Engineering*, 6, 77–98.

Ghosh, S., Mondal, S., & Ghosh, B. (2014, February). A comparative study of breast cancer detection based on SVM and MLP BPN classifier. In *2014 first international conference on automation, control, energy and systems (ACES)* (pp. 1–4). IEEE.

Han, X., Zhong, Y., Cao, L., & Zhang, L. (2017). Pre-trained alexnet architecture with pyramid pooling and supervision for high spatial resolution remote sensing image scene classification. *Remote Sensing*, 9(8), 848.

Hassan, A. M., & El-Shenawee, M. (2011). Review of electromagnetic techniques for breast cancer detection. *IEEE Reviews in Biomedical Engineering*, 4, 103–118.

Jha, A., Verma, G., Khan, Y., Mehmood, Q., Rebholz-Schuhmann, D., & Sahay, R. (2018, December). Deep convolution neural network model to predict relapse in breast cancer. In *2018 17th IEEE international conference on ML and applications (ICMLA)* (pp. 351–358). IEEE.

Khuriwal, N., & Mishra, N. (2018, October). Breast cancer diagnosis using DL algorithm. In *2018 International Conference on Advances in Computing, Communication Control and Networking (ICACCCN)* (pp. 98–103). IEEE.

Medjahed, S. A., Saadi, T. A., &Benyettou, A. (2013). Breast cancer diagnosis by using k-nearest neighbor with different distances and classification rules. *International Journal of Computer Applications*, 62(1), (pp. 1–15).

Nawaz, W., Ahmed, S., Tahir, A., & Khan, H. A. (2018). Classification of breast cancer histology images using alexnet. In *Image Analysis and Recognition: 15th International Conference, ICIAR 2018, Póvoa de Varzim, Portugal, June 27–29, 2018, Proceedings 15* (pp. 869–876). Springer International Publishing.

Oliveira, J. E., Gueld, M. O., Araújo, A. D. A., Ott, B., &Deserno, T. M. (2008, March). Toward a standard reference database for computer-aided mammography. In *Medical imaging 2008: Computer-aided diagnosis* (Vol. 6915, pp. 606–614). SPIE.

Rathi, M., & Pareek, V. (2016). Hybrid approach to predict breast cancer using ML techniques. *International Journal of Computer Science Engineering*, 5(3), 125–136.

Shravya, C., Pravalika, K., & Subhani, S. (2019). Prediction of breast cancer using supervised ML techniques. *International Journal of Innovative Technology and Exploring Engineering (IJITEE)*, 8(6), 1106–1110.

Sivapriya, J., Kumar, A., Sai, S. S., & Sriram, S. (2019). Breast cancer prediction using ML. *International Journal of Recent Technology and Engineering (IJRTE)*, 8(4), 4879–4881.

Soni, B., Bora, A., Ghosh, A., & Reddy, A. (2019). RFSVM: A novel classification technique for breast cancer diagnosis. *International Journal of Innovative Technology and Exploring Engineering (IJITEE)*, 8(12), 3295–3305.

Sornam, M., Muthusubash, K., & Vanitha, V. (2017, December). A survey on image classification and activity recognition using deep convolutional neural network architecture. In *2017 ninth international conference on advanced computing (ICoAC)* (pp. 121–126). IEEE.

Srirambabu, S., Kumar, V. S., & Murugan, B. S. (2010). Classification of breast cancer using SVM classifier technique. *International Journal of Advanced Research in Computer Science*, 1(4), 261–263.

Sun, Y. S., Zhao, Z., Yang, Z. N., Xu, F., Lu, H. J., Zhu, Z. Y., ... & Zhu, H. P. (2017). Risk factors and preventions of breast cancer. *International Journal of Biological Sciences*, 13(11), 1387.

Zheng, B., Yoon, S. W., & Lam, S. S. (2014). Breast cancer diagnosis based on feature extraction using a hybrid of K-means and support vector machine algorithms. *Expert Systems with Applications*, 41(4), 1476–1482.

Chapter 13

Brain tumor identification using DenseNet-201 deep learning model

Soubhagya Das and Bidush Kumar Sahoo

13.1 INTRODUCTION

An abnormal proliferation of cells within the brain is referred to as a BT. These tumors can vary in size, location, and aggressiveness, posing a significant health risk due to their potential to im-pair cognitive abilities, motor functions, and even vital organ systems. Tumors, on the other hand, represent an aberrant proliferation of fibrous tissue within the brain, growing unchecked and disrupting its normal function [DeAngelis, 2001]. The diagnosis and treatment of BTs are critical endeavors, often involving a combination of imaging techniques, like MRI and CT scans, along with surgical interventions, radiation therapy, and chemotherapy [Badža & Barjaktarović, 2020].

Technological progress has transformed medical imaging, with ML and DL methods becoming essential for improving diagnostic precision and treatment strategies. These techniques aid in the automated examination of several medical pictures to identify and diagnose illnesses such as cancer, neovascularization, and COVID-19. Their studies have showed promise in enhancing accuracy, especially in early illness identification, which is crucial for efficient treatment [Işın et al., 2016]. BTs provide a substantial health hazard, requiring timely identification for the best treatment results. DL, especially in the examination of MRI data, has potential in precisely identifying brain cancers in their early stages. Many studies have shown the strong precision of DL algorithms in this area; however, further study is required to thoroughly investigate the capabilities of pre-trained networks in detecting BTs. DL is crucial in detecting BTs by using sophisticated computer methods to examine intricate patterns in medical imaging data, including MRI images. CNNs and similar algorithms may effectively detect minor irregularities that suggest the presence of BTs with great precision. DL models may be trained on vast datasets to differentiate between normal brain tissue and diseased characteristics, allowing for early and accurate tumor diagnosis. Furthermore, DL methods have the capability to automate and scale, aiding healthcare practitioners in interpreting imaging data quickly and reliably. By continuously refining and validating DL approaches, they greatly enhance

DOI: 10.1201/9781003589716-13

the efficiency and efficacy of BT identification, thereby improving patient care and outcomes [Al-Galal et al., 2021, Hashemzehi et al., 2020].

13.1.1 BT identification using ML

BTs are among the deadliest diseases and require early detection for treatment. Traditional diagnostic methods like MRI and CT scans depend on radiologists, making them subjective, time-consuming, and error-prone. ML has revolutionized BT identification in healthcare. ML provides faster, more accurate, and automated tumor detection, classification, and prognosis using advanced algorithms to analyze complex datasets, improving patient care. ML can identify BTs by processing and analyzing large amounts of medical data. Advanced imaging techniques produce complex data that is difficult to interpret manually. ML algorithms, especially DL ones, can spot subtle patterns and anomalies in such data. CNNs, designed to process images, are leading this revolution. CNNs automatically detect tumor boundaries, shape, size, and texture from MRI or CT scans. These features help distinguish gliomas, meningiomas, and pituitary tumors from normal brain tissue.

Pre-processing is essential for BT detection by ML models. Medical images with noise, artifacts, and varying intensities can obscure important information. Noise reduction, histogram equalization, and contrast enhancement improve image clarity and feature extraction. Another important pre-processing step is tumor segmentation from surrounding brain structures. For precise tumor boundaries, U-Net and fully convolutional networks are effective segmentation methods. These methods let ML models focus on relevant areas, improving diagnostic accuracy. ML models can combine imaging, demographic, genetic, and clinical data to understand the tumor. Multimodal approaches enhance prediction accuracy and treatment customization. Combining radiological data with genetic markers can predict tumor aggressiveness or therapy response. This capability supports precision medicine, which tailor's treatments to each patient.

Explorers are using unsupervised learning as well as supervised learning. These methods can cluster patients by tumor characteristics or find new biomarkers in datasets. Medical research is often hampered by limited labeled data. Semi-supervised and reinforcement learning methods address this issue. Transfer learning, where pre-trained models are adapted to specific tasks, reduces computational costs and speeds model deployment. ML in BT identification faces several obstacles. High-quality, annotated medical datasets, and imaging protocol standardization across institutions are major limitations. To gain patient and practitioner trust, ethical issues like data privacy, algorithmic bias, and ML prediction interpretability must be addressed. Explainable AI (XAI) offers human-readable prediction justifications for ML decision-making, a promising solution. ML integration into clinical workflows needs rigorous validation and regulatory approval.

Computer scientists, radiologists, and oncologists must work together to apply technology in real life. Continuous advances in computing power, algorithm design, and dataset availability will drive innovation in oncology, making ML essential. ML has improved BT detection accuracy, efficiency, and scalability. Its ability to analyze complex medical data and provide actionable insights could improve early detection, prognosis, and patient outcomes. ML integration into clinical practice will transform BT management as technology advances.

13.1.2 BT identification using DL

DL, a subset of ML, has transformed BT detection in medical imaging. DL improves tumor detection accuracy, speed, and automation by learning and extracting complicated information from raw data. CNNs, a common DL architecture, can now evaluate brain imaging data like MRI and CT scans with unparalleled precision. These models identify subtle patterns, textures, and anomalies in medical images without manual feature extraction, aiding diagnosis and therapy planning. Image categorization is a major DLBT identification application. CNNs can classify brain scans as normal or tumor-affected and identify between gliomas, meningiomas, and pituitary adenomas. Various tumor forms require various treatments; therefore, these classifications help guide therapeutic decisions. CNN-based models like ResNet, VGGNet, and DenseNet accurately identify tumor areas in big datasets with low false-positive and false-negative rates.

DL has improved BT segmentation, another crucial step. Correct segmentation is needed to define tumor boundaries, measure size, and evaluate growth and treatment response. By learning spatial hierarchies and contextual interactions, U-Net, a popular DL architecture for medical picture segmentation, can identify tumor regions. 3D U-Net and Attention U-Net use three-dimensional data and attention mechanisms to focus on the most important sections of images to improve performance. These methods segment tumors accurately even in low-contrast or irregular tumors. Comprehensive tumor analysis requires multimodal data integration, which DL enables. DL models can provide a complete tumor picture by incorporating radiological, genomic, histological, and clinical data. This method is useful for predicting tumor grade, aggressiveness, and patient outcomes. Hybrid CNN-RNN models can assess temporal patterns in longitudinal imaging investigations to track tumor growth. The versatility of DL models like transfer learning has boosted their use in BT identification. Transfer learning fine-tunes medical imaging algorithms pre-trained on big, generic datasets like ImageNet. This minimizes the need for large, expensive medical datasets and speeds up diagnostic tool development. DL in BT identification is challenging despite its transformative potential. Annotated, high-quality datasets are still a constraint. DL model interpretability and openness are also crucial for clinical uptake. Explainable AI (XAI) approaches are being developed to visualize and explain the model's decision-making process, boosting medical professional trust.

DL has revolutionized BT identification with its classification, segmentation, and data integration capabilities. DL will enable early diagnosis, tailored treatment planning, and better patient outcomes as research and technology advance.

13.1.3 BT identification using ensemble learning

ML ensemble learning, which integrates model predictions, is effective at BT identification. The capabilities of individual models are used to develop a more robust and accurate forecasting system. Ensemble learning can improve diagnostic accuracy, reduce false positives and negatives, and help clinical decision-making in medical imaging, where precision and reliability are crucial. Ensemble learning is a powerful method for BT identification and characterization because it integrates multiple models. Ensemble learning uses decision trees, SVMs, and neural networks to improve performance. Bagging, boosting, and stacking are ensemble learning methods. Bootstrap aggregation, or bagging, trains multiple models on various data subsets and averages their predictions. Brain tumor classification using Random Forest, a bagging-based method, is popular. It works well with noisy or imbalanced datasets because it combines decision tree outputs to improve prediction stability and reduce volatility.

However, boosting trains models progressively to repair earlier errors. Tumor identification is often improved with AdaBoost, GBM, and XGBoost. Boosting strategies steadily improve model performance, making them ideal for complicated datasets like brain imaging data. Gradient Boosting can learn from misclassified instances to accurately categorize brain cancers into gliomas, meningiomas, and pituitary adenomas. A meta-model learns how to incorporate the predictions of numerous base models in stacking, a more advanced ensemble method. Stacking helps BT identification by merging the results of diverse models like CNNs for image analysis and Random Forests for clinical data interpretation. This method ensures that the ensemble combines the strengths of different models for more precise and trustworthy forecasts. Brain tumor segmentation is also improved by ensemble learning. Precision segmentation defines tumor boundaries and aids treatment planning. Ensemble approaches improve tumor border identification by merging segmentation models like U-Net and DeepLab. These methods have reduced over- and under-segmentation in MRI data, yielding better results than individual models.

Ensemble learning has drawbacks like computational complexity and overfitting. Modern computational resources and methods like cross-validation and regularization have reduced these concerns. Clinical applications still struggle with interpretability, but Explainable AI (XAI) approaches and ensemble learning can make predictions transparent and reliable for medical practitioners. Ensemble learning is a potent BT diagnosis method that combines numerous models for high accuracy and dependability. For early diagnosis and individualized therapy, its classification, segmentation, and multimodal data integration applications are essential. Ensemble approaches will become more important in clinical operations as research improves patient outcomes and medical imaging technologies.

13.1.4 BT identification using explainable AI (XAI)

XAI BT diagnosis combines DL accuracy with clinical interpretability, advancing medical diagnostics. Traditional AI models, particularly DL models, are often "black boxes," giving accurate findings but providing little insight into decision-making. In medical applications, where confidence and clinical acceptance depend on prediction reasoning, this lack of transparency might be a big drawback. By revealing how AI models work, XAI helps physicians comprehend, interpret, and evaluate predictions, which is crucial for brain tumor diagnosis. Identifying brain cancers in MRI or CT imaging, detecting tumor types, and determining tumor borders for treatment planning are hard jobs. XAI methods including saliency maps, Grad-CAM, and SHAP are being used to make AI predictions more understandable. These methods show doctors which aspects of the input image influence the AI model's prediction, allowing them to check if the model is targeting tumor locations. Grad-CAM may identify tumor spots in an MRI scan that the model links with malignancy, ensuring that the AI system is assessing the image as a person would.

Using XAI, BT diagnosis systems can find biases in training data or prediction errors to improve model generalization and reliability. XAI approaches that show a model focusing on non-tumor regions or visual artifacts indicate the need for better data pre-processing or augmentation. Our continuous improvement process delivers accurate and resilient models across varied patient groups and imaging modalities. The clinical decision-making benefit of XAI in BT identification is also important. Traditional AI models produce binary outputs ("tumor" or "no tumor"), whereas XAI-enhanced systems can provide probabilistic forecasts and thorough explanations, allowing physicians to combine AI insights with their experience. This collaboration is especially useful in uncertain instances, when the AI may detect small irregularities that the human sight misses. XAI may also validate or dispute AI predictions, ensuring informed, accurate, and transparent healthcare judgments. Explainable AI could improve oncology tailored medication. Using imaging data and patient-specific information including genetic profiles and clinical histories, XAI algorithms can predict tumor behavior, treatment response, and prognosis. An XAI model could detect glioblastomas and explain how their characteristics relate to patient-specific risk factors, directing surgery, chemotherapy, or radiotherapy. XAI in BT identification has drawbacks. XAI interpretability must be extensively verified to assure accurate and therapeutically useful interpretations. Research continues on creating XAI systems that balance interpretability and computational efficiency. However, XAI's ability to improve BT diagnostics' confidence, transparency, and accuracy could alter current healthcare. XAI assures that BT recognition technology improves diagnosis, treatment, and outcomes by bridging the gap between cutting-edge AI and clinical experience.

This research intends to explore the capabilities of DenseNet-201, a widely used pre-trained neural network, for identifying brain cancers in MRI data. The project aims to enhance BT identification accuracy and efficiency by using DL methods via thorough examination and analysis of its performance.

The chapter is divided into sections: Section 13.2 reviews relevant work, Section 13.3 describes the technique, Section 13.4 presents the findings, and Section 13.5 includes concluding remarks.

13.2 LITERATURE REVIEW

This segment explores the diverse range of uses for ML and DL, including the interpretation of medical images and the investigation of infectious brain lesions. In recent decades, there has been considerable interest and scholarly attention directed toward medical image analysis owing to its multifaceted utility in the healthcare sector, particularly in the areas of patient examination and diagnosis.

[Razi et al., 2023] examined the important medical issue of identifying BTs by automatically categorizing MRI images. An advanced DL approach was used, using 8 iterations of the EfficientNet (B0-B7) model. It achieved an accuracy of 99.67% and a training accuracy of 99.71%, showing exceptional achievements. The improved performance was achieved with faster computing compared to the classic CNN technique. [Havaei et al., 2017] introduced the "BraTS network," a deep CNN architecture designed for BT segmentation using both 2D and 3D data. Their model surpassed others on the BraTS datasets by using multi-scale processing to include local and global contextual information. [Seetha & Raja, 2018] developed a sophisticated convolutional neural network system designed to automatically identify and classify brain cancers. The method utilized Fuzzy C-Means with feature extraction mechanism. The characteristics are then inputted into SVM and DNN classifiers. The system had an accuracy level of 97.5% according to the findings. [Bashkandi et al., 2023] addressed the critical medical challenge of automated MRI picture classification to identify brain cancers. DL was carried out using eight distinct versions of the EfficientNet model. It's B7 version model performed training accuracy of 99.71%. [Sasikala & Kumaravel, 2008] introduced a genetic algorithm based on ANN for feature selection to reduce the dimensionality of wavelet feature sets. This algorithm successfully identified only four out of twenty-nine features and achieved an impressive accuracy of 98% by utilizing only the selected features. The neutrosophic set expert maximal fuzzy-sure method was employed by [Özyurt et al., 2019] for segmenting BTs, followed by feature extraction using CNN on the resulting images. Following this, SVM classifiers were utilized to train on these features in order to categorize the tumors as either benign or malignant. The strategy implemented yielded a mean success rate of 95.62%. [Kumar et al., 2021] conducted a comprehensive study exploring various ML and DL techniques to segment and detect BTs. Four basic ML algorithms were utilized in this study. Classic SVMs exhibited the highest level of classification accuracy, achieving 92.4%. Moreover, a five-layer custom CNN architecture was developed by the researchers with the specific purpose of identifying BTs in MRI scans. The implementation demonstrated an impressive accuracy rate of 97.2%. [Khan et al., 2021] suggested a methodology that employs the VGG19 model and K-means clustering for the classification and segmentation of brain malignancies in MRI images. The methodology utilized involved segmenting the input MR modality and pre-processing intensities through statistical normalization. The approach demonstrated a precision rate of 94% overall. [Bhatele & Bhadauria, 2022] developed ML model to differentiate between high- and low-grade glioma images. Researchers performed classification accuracies of 90% and 95% for brain malignancies and other central nervous system components using an extreme gradient boosting model. [Kang et al., 2021] who employed a deep CNN feature ensemble along with ML classifiers. The experiments conducted by the researchers' involved datasets of various sizes. When compared to other classifiers, the SVM utilizing a radial basis mechanism exhibited the best performance. Grampurohit et al. developed a CNN model to classify 253 BT images into two categories: 155 tumors and 98 non-tumorous images. The model utilized the VGGNet architecture. To address the problem of overfitting, strategies like data augmentation and pre-processing were utilized to enhance the variety of the samples. When evaluating a single dataset, the custom CNN model showed a validation accuracy of 86%, while VGGNet achieved an impressive accuracy of 97%. [Mehrotra et al., 2020] conducted a study investigating different DL-based pre-trained CNN algorithms specifically created to distinguish between benign and malignant brain lesions. In order to achieve their objectives, the researchers explored various optimizers, such as Adam, RMSprop, and stochastic gradient descent. The study findings indicate that AlexNet exhibited outstanding performance in medical imaging tasks following optimal tuning. [Kamnitsas et al., 2017] introduced "DeepMedic," a three-dimensional CNN architecture. The model developed by the researchers integrated contextual information and multi-scale inputs through conditional random fields. The methodology described above demonstrated effective results when implemented on the BraTS 2013 and 2014 datasets. [Sangeetha et al., 2020] examined the classification performance of different

CNN architectures in their research. ResNet-50 exhibited superior performance compared to GoogleNet and VGGNets, achieving an accuracy rate of 96.50%, surpassing the 93.45% and 89.33% achieved by the other models, respectively. Furthermore, ResNet-50 exhibited a 10% improvement in accuracy when compared to VGGNet and GoogleNet, while also showing a 10% decrease in data processing time.

While the above studies outlined various successful methodologies for BT identification using DL approaches, there are some limitations to consider. These include potential biases in the datasets used, limited generalizability to diverse patient populations or imaging modalities, and challenges related to interpretability and explainability of the DL models, which are crucial in clinical decision-making.

13.2.1 Literature survey with advantages and limitations

Table 13.1 represents the advantages and limitations of various models presented in the literature survey.

13.3 MATERIALS AND METHODS

The study aims to use the DenseNet-201 model to boost accuracy in picture classification, minimizing the need for human interpretation and enabling future system improvements. The procedure includes many stages: acquiring input images, pre-processing, performing erosion and dilation, and classifying utilizing DenseNet-201 techniques, as shown in Figure 13.1.

13.3.1 DenseNet-201

We used the DenseNet-201 DL model, which consists of 201 layers. The condensed network is used, offering efficient and easily trainable methods [Huang et al., 2017]. This is due to the capability of many layers recycling characteristics, which boosts variance in future layer input and improves performance. It surpassed ImageNet and CIFAR-100 on many datasets. DenseNet-201 utilizes feed-forward connections to connect layers. The feature maps of each layer are based on the feature maps of the preceding layer. This network assimilates information from all previous levels. Each connection from the preceding layer is included into the DenseNet-201 model. DenseNet addresses the vanishing gradient issue by aggregating feature maps from all previous layers. It improves feature propagation and reduces the number of parameters. Due to its compact architecture, it has fewer parameters and less processing resources than other deep networks of equal depth, enabling faster convergence and training. In BT identification, DenseNet-201 is advantageous. The ability to propagate characteristics

Table 13.1 Advantages and limitations of various models for BT identification

Study	*Methodology*	*Advantages*	*Limitations*
[Razi et al., 2023]	EfficientNet (B0-B7) model for MRI categorization	Exceptional accuracy (99.67%), faster computation compared to classic CNNs	High dependency on EfficientNet structure; limited applicability to other imaging modalities
[Havaei et al., 2017]	BraTS network (deep CNN) for tumor segmentation using 2D and 3D data	Multi-scale processing incorporates local and global context; excellent segmentation results	Computationally intensive; requires extensive labeled datasets
[Seetha & Raja, 2018]	CNN with Fuzzy C-Means and SVM/DNN classifiers	High accuracy (97.5%); hybrid approach enhances feature representation	Complexity in combining techniques; potential overfitting
[Bashkandi et al., 2023]	EfficientNet model variants for automated MRI classification	Outstanding accuracy (99.71%); efficient computational performance	Requires substantial computational resources for large EfficientNet models
[Sasikala & Kumaravel, 2008]	Genetic algorithm with ANN for wavelet feature selection	Reduced dimensionality with minimal loss of information; 98% accuracy	Limited to specific feature extraction methods; complexity in tuning genetic algorithm
[Özyurt et al., 2019]	Neutrosophic set method with CNN and SVM for segmentation and classification	Mean success rate of 95.62%; integration of fuzzy logic improves segmentation	Dependent on expert-tuned parameters; challenges with generalizability
[Kumar et al., 2021]	Custom CNN with SVM and ML-based segmentation	High classification accuracy (97.2%); versatile ML and DL approach	Limited dataset diversity; challenges in integrating multimodal data
[Khan et al., 2021]	VGG19 and K-means clustering for MRI segmentation and classification	Combines clustering and DL for robust pre-processing; achieved 94% precision	Data normalization constraints; sensitive to initial clustering parameters
[Bhatele & Bhadauria, 2022]	Extreme Gradient Boosting for glioma classification	Effective high/low-grade glioma differentiation; achieves up to 95% accuracy	Requires large, balanced datasets; less interpretability compared to simpler models

(*Continued*)

Table 13.1 (Continued)

Study	*Methodology*	*Advantages*	*Limitations*
[Kang et al., 2021]	Deep CNN feature ensemble with SVM classifiers	Superior performance with SVM radial basis function; effective on varied datasets	Computational cost for feature ensemble; risk of overfitting with high-dimensional data
[Grampurohit et al., 2020]	VGGNet with data augmentation for classification of tumorous and non-tumorous images	Reduced overfitting through augmentation; achieved 97% accuracy with VGGNet	Relatively lower validation accuracy for custom CNN (86%); dependent on augmentation quality
[Mehrotra et al., 2020]	Pre-trained CNNs (e.g., AlexNet) with optimizers (Adam, RMSprop, SGD)	Outstanding performance post-optimization; broad utility across imaging tasks	Requires careful optimizer tuning; limited validation across diverse imaging datasets
[Kamnitsas et al., 2017]	DeepMedic 3D CNN with conditional random fields	Integrates multi-scale inputs; effective on BraTS datasets	Complex architecture increases computational load; limited testing beyond BraTS datasets
[Sangeetha et al., 2020]	Comparative analysis of CNN architectures (ResNet-50, VGGNet, GoogleNet)	ResNet-50 achieves highest accuracy (96.5%); faster processing time	Dependency on specific architecture (ResNet); less improvement with smaller datasets

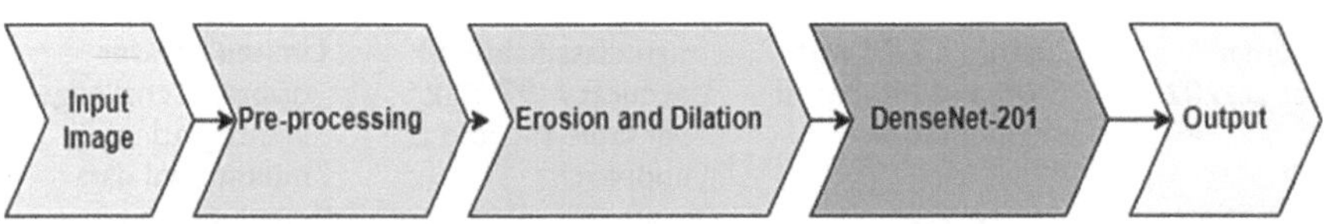

Figure 13.1 Block Diagram of BT Identification.

across layers preserves subtle malignancy patterns, resulting in very accurate classifications. This design works well in medical imaging, since datasets and annotations are generally small. It generalizes well in low-data contexts due to its feature reuse, making it appropriate for resource-constrained healthcare applications. Due to its efficiency, the design can be used in real-time diagnostic systems for clinical accuracy and speed.

It may combine features from all layers, making it versatile for image classification and segmentation. In BT studies, precise tumor boundary delineation is essential for therapy planning. The model's feature-sharing technique reduces depth and complexity, improving accuracy and simplifying network design for high-performance models. It significantly outperforms standard architectures in discriminating normal and malignant tissues, classifying tumor kinds, and locating and extending tumors. It improves DL efficiency, accuracy, and scalability. Its revolutionary architecture has raised the bar in many applications, including medical imaging, where it can diagnose and analyze brain cancers. The architecture of DenseNet201 is seen in Figure 13.2.

13.3.2 Dataset and pre-processing

The dataset utilized for implementation comprises 2200 brain MRI pictures categorized into two folders: "yes" with 1100 tumorous brain MRI images and "no" with 1100 non-tumorous brain MRI images [Brain Tumor Dataset, 2024]. The photos were standardized to a shape of (240, 240, 3) during data pre-processing to maintain uniformity in size and orientation. This uniformity allows them to be inputted into the neural network consistently. Morphological closure in MRI image processing involves dilatation followed by erosion to enhance the clarity of tumor areas. Dilation enlarges and links cancerous pixels, whereas erosion refines them by removing noise. This technique leads to clearer and more cohesive tumor representations, improving the precision of analysis and diagnosis.

Normalization is an essential pre-processing procedure that entails modifying the range of pixel intensity values in photographs. This method, also known as histogram stretching or contrast stretching, is very beneficial for photographs with low contrast. The pixel values were normalized to a range of 0 to 1 for optimal display. The process involved multiplying all pixel values by the maximum pixel value of 255. Sample BT pictures are depicted in Figure 13.3. We divided our dataset into 80% training and 20% testing portions and shuffled it 100 times to enhance model generalization by reducing variance and loss in order to be compatible with the DenseNet-201 model.

13.4 EXPERIMENTAL DESIGN AND RESULT ANALYSIS

Python 3.6 and PyCharm, a robust and versatile ML programming environment, were used to create the BT diagnosis system. The framework was built for Windows 11, ensuring compatibility with new software and hardware. The testing environment had a high-performance GPU with 8 GB of RAM to handle DL models' computationally expensive activities, especially for training huge networks like DenseNet-201. Keras, a TensorFlow-based high-level neural networks API, was utilized to create the DL model. An easy-to-use interface allows you create, train, and evaluate DL models. Python packages

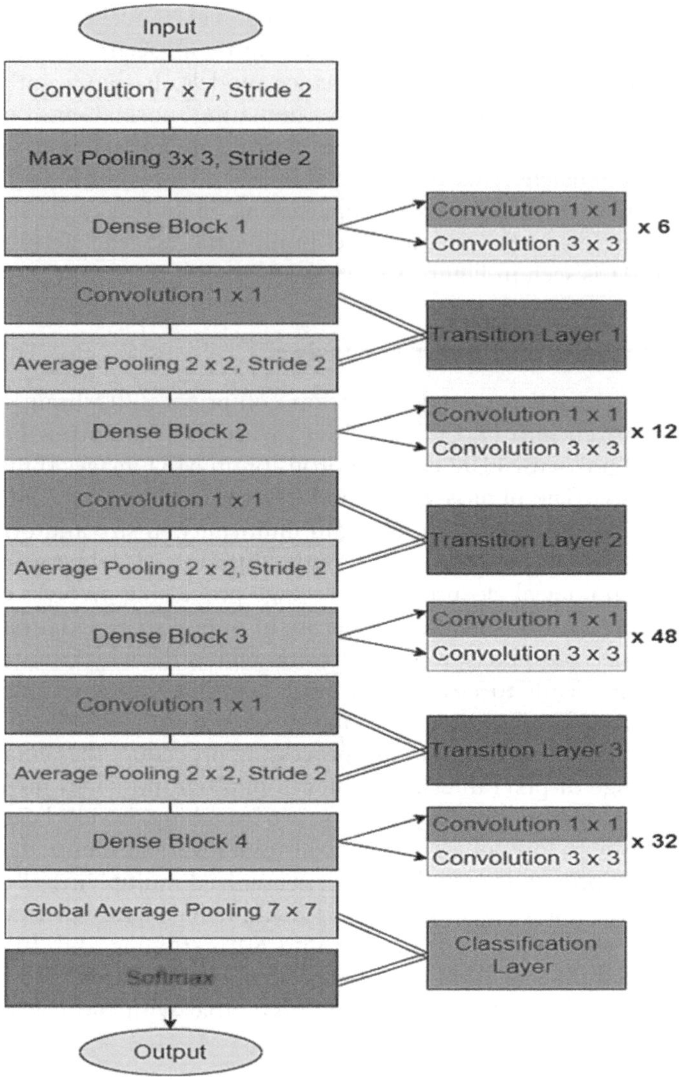

Figure 13.2 DenseNet-201.

Scipy, Numpy, and OpenCV are used for pre-processing and data manipulation. Scipy and Numpy solve numerical tasks like matrix manipulations and create mathematical functions needed to process medical pictures and optimize model parameters. Image pre-processing operations including scaling, normalization, and augmentation using OpenCV, a computer vision toolkit, improve the model's generalization across varied image datasets.

Hyperparameters were tuned across 25 epochs to balance training time and model performance in DL networks. The commonly used Adam

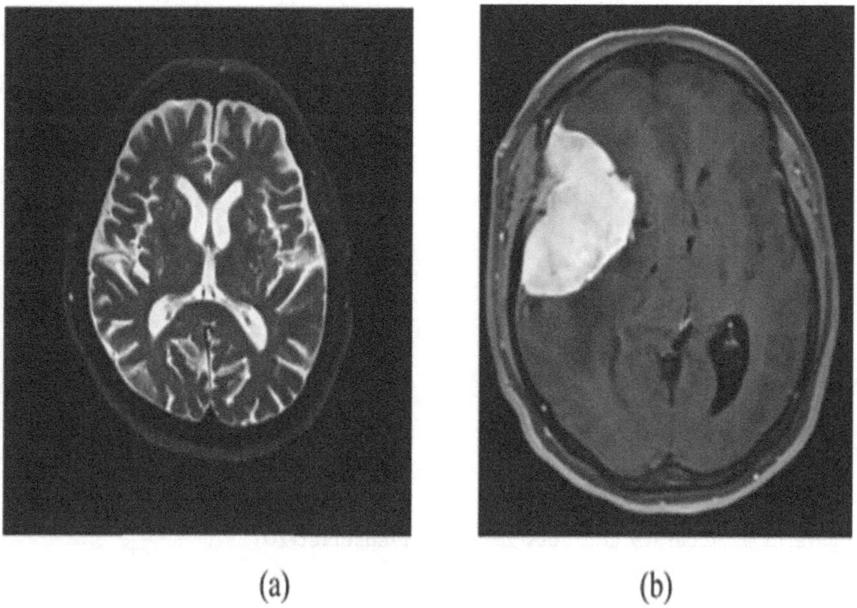

Figure 13.3 (a) Healthy brain (b) BT images.

optimizer combined the benefits of AdaGrad and RMSProp optimizers to help the model converge rapidly and efficiently. Adam was chosen for processing noisy data like medical photos with artifacts or inconsistencies due to its adjustable learning rate. A low learning rate of 0.0001 prevents overfitting and provides stable convergence by gradually updating the model's weights. The DenseNet-201 model's hidden layers used ReLU (Rectified Linear Unit) activation functions to introduce non-linearity, allowing the network to capture complicated input data patterns and attributes for successful tumor diagnosis. Figure 13.4 shows accuracy and loss curves used to monitor model performance during training. The accuracy curve shows how well the model matches labels, while the loss curve shows the gap between expected outputs and ground truth. These curves show the model's learning behavior over time. A consistent increase in accuracy and decrease in loss imply successful learning and model modification. With its sophisticated dense connection architecture, the DenseNet-201 model was able to propagate features efficiently and avoid the vanishing gradient problem, which plagues deep networks. The refined testing methodology and parameter selections assure that the framework can detect BTs efficiently and reliably for real-time medical applications.

The DenseNet-201 model's performance was assessed based on recall, precision, F1-Score, sensitivity, and specificity, with the findings displayed in Figure 13.5. The confusion matrix for DenseNet-201 is seen in Figure 13.6. The performance metrics were calculated using the following formulae.

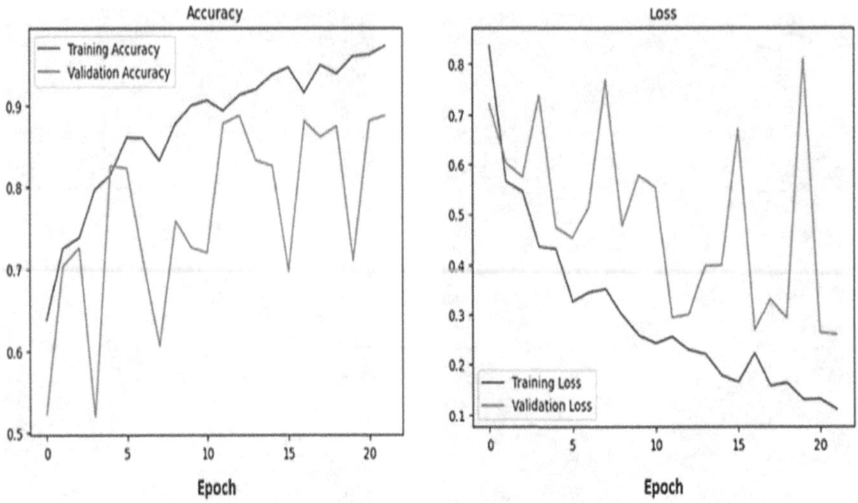

Figure 13.4 Accuracy and Loss graph for DenseNet-201.

1. $\text{Precision} = \dfrac{\text{True Positive}(\text{TP})}{\text{True Positive} + \text{False Positive}(\text{FP})}$
2. $\text{Sensitivity or Recall} = \dfrac{\text{True Positive}}{\text{True Positive} + \text{False Negative}(\text{FN})}$
3. $\text{F1} - \text{Score} = \dfrac{2 \times (\text{Precision} \times \text{Recall})}{(\text{Precision} + \text{Recall})}$
4. $\text{Specificity} = \dfrac{\text{True Negative}(\text{TN})}{\text{True Negative} + \text{False Positive}}$
5. $\text{Accuracy} = \dfrac{\text{TP} + \text{TN}}{\text{TP} + \text{TN} + \text{FP} + \text{FN}}$

Precision: The percentage of model positive predictions that are true. High precision reduces false positives. Recall: It evaluates the model's ability to recognize all true positives. High recall reduces false negatives.

F1-Score: The harmonic mean of Precision and Recall provides a single score to assess their balance. Specificity: The model's ability to prevent false positives is measured by the percentage of true negatives properly detected out of all negatives.

Accuracy: It determines model correctness by considering genuine positives and negatives.

The confusion matrix details the model's BT picture classification into "Healthy" and "Tumor." TP, FP, FN, and TN are the matrix's four main

Class Name	Precision	1-Precision	Recall	1-Recall	f1-score
Healthy	0.9773	0.0227	0.9862	0.0138	0.9817
Tumor	0.9864	0.0136	0.9775	0.0225	0.9819
Accuracy			0.9818		
Misclassification Rate			0.0182		
Macro-F1			0.9818		
Weighted-F1			0.9818		

Figure 13.5 Performance measures for DenseNet-201 method.

values. In this situation, the model accurately predicted 215 "Healthy" cases as "Healthy" and 217 "Tumor" instances as "Tumor." However, the model misclassified three "Healthy" cases as "Tumor" and five "Tumor" cases as "Healthy." These parameters are utilized to construct numerous key model accuracy and reliability metrics. The model's accuracy is the percentage of right "Healthy" and "Tumor" predictions out of all forecasts, which is 98.18%. Precision, which shows how many "Tumor" cases were correctly predicted, is 98.64%, indicating that the model is accurate. The model diagnoses most tumor instances with a recall of 97.75% for "Tumor," which evaluates sensitivity. Specifically, 97.73% of actual "Healthy" cases were accurately recognized as "Healthy," showing that the model can also identify non-tumor patients. Finally, the F1-Score, which balances precision and recall, is 98.19%, suggesting the model's ability to balance false positives and negatives and perform well. The confusion matrix and derived metrics show that the model classifies BT images accurately, making it reliable for medical diagnosis.

TARGET / OUTPUT	Healthy	Tumor	SUM
Healthy	215 48.86%	5 1.14%	220 97.73% 2.27%
Tumor	3 0.68%	217 49.32%	220 98.64% 1.36%
SUM	218 98.62% 1.38%	222 97.75% 2.25%	432 / 440 98.18% 1.82%

Figure 13.6 DenseNet-201 Confusion matrix.

13.5 CONCLUSION AND FUTURE SCOPE

Early identification of BTs is essential for improving treatment results. BT identification technologies help medical practitioners find tumors early, facilitating prompt intervention and treatment. Timely identification enhances the likelihood of effective therapy, higher survival rates, and better quality of life for those with BTs. We utilized several MRI BT images in this study to detect early-stage brain cancers. DL approaches can also assist in categorization and detection. The proposed DenseNet-201 model has an accuracy rating of 98.18%. Identifying BTs has several benefits such as early detection, better treatment results, improved patient care, decreased need for unneeded treatments, support for research and progress, increased public knowledge and education, and optimized resource allocation. Future study might involve investigating sophisticated CNN structures, including more modalities for a more thorough analysis, and creating techniques for real-time use in clinical environments. The enhancements are intended to improve the strength and practicality of the suggested system for identifying brain cancers.

REFERENCES

Al-Galal, S. A. Y., Alshaikhli, I. F. T., & Abdulrazzaq, M. M. (2021). MRI brain tumor medical images analysis using deep learning techniques: a systematic review. *Health and Technology*, 11, 267–282.

Badža, M. M., & Barjaktarović, M. Č. (2020). Classification of brain tumors from MRI images using a convolutional neural network. *Applied Sciences*, 10(6), 1999.

Bashkandi, A. H., Sadoughi, K., Aflaki, F., Alkhazaleh, H. A., Mohammadi, H., & Jimenez, G. (2023). Combination of political optimizer, particle swarm optimizer, and convolutional neural network for brain tumor detection. *Biomedical Signal Processing and Control*, 81, 104434.

Bhatele, K. R., & Bhadauria, S. S. (2022). Machine learning application in Glioma classification: review and comparison analysis. *Archives of Computational Methods in Engineering*, 29(1), 247–274.

Brain Tumor Dataset (2024) Online Available: https://www.kaggle.com/jakeshbohaju/brain-tumor [accessed on 12 April 2024].

DeAngelis, L. M. (2001). Brain tumors. *New England Journal of Medicine*, 344(2), 114–123.

Grampurohit, S., Shalavadi, V., Dhotargavi, V. R., Kudari, M., &Jolad, S. (2020, October). Brain tumor detection using deep learning models. In *2020 IEEE India Council International Subsections Conference (INDISCON)* (pp. 129–134). IEEE.

Hashemzehi, R., Mahdavi, S. J. S., Kheirabadi, M., & Kamel, S. R. (2020). Detection of brain tumors from MRI images base on deep learning using hybrid model CNN and NADE. *Biocybernetics and Biomedical Engineering*, 40(3), 1225–1232.

Havaei, M., Davy, A., Warde-Farley, D., Biard, A., Courville, A., Bengio, Y., ... & Larochelle, H. (2017). Brain tumor segmentation with deep neural networks. *Medical Image Analysis*, 35, 18–31.

Huang, G., Liu, Z., Van Der Maaten, L., & Weinberger, K.Q. (2017). Densely connected convolutional networks. In *Proceedings of the IEEE Conference on Computer Vision and Pattern Recognition 2017* (pp. 4700–4708).

Işın, A., Direkoğlu, C., & Şah, M. (2016). Review of MRI-based brain tumor image segmentation using deep learning methods. *Procedia Computer Science*, 102, 317–324.

Kamnitsas, K., Ledig, C., Newcombe, V. F., Simpson, J. P., Kane, A. D., Menon, D. K., ... & Glocker, B. (2017). Efficient multi-scale 3D CNN with fully connected CRF for accurate brain lesion segmentation. *Medical Image Analysis*, 36, 61–78.

Kang, J., Ullah, Z., & Gwak, J. (2021). MRI-based brain tumor classification using ensemble of deep features and machine learning classifiers. *Sensors*, 21(6), 2222.

Khan, A. R., Khan, S., Harouni, M., Abbasi, R., Iqbal, S., & Mehmood, Z. (2021). Brain tumor segmentation using K-means clustering and deep learning with synthetic data augmentation for classification. *Microscopy Research and Technique*, 84(7), 1389–1399.

Kumar, G., Kumar, P., & Kumar, D. (2021, December). Brain tumor detection using convolutional neural network. In *2021 IEEE International Conference on Mobile Networks and Wireless Communications (ICMNWC)* (pp. 1–6). IEEE.

Mehrotra, R., Ansari, M. A., Agrawal, R., & Anand, R. S. (2020). A transfer learning approach for AI-based classification of brain tumors. *Machine Learning with Applications*, 2, 100003.

Özyurt, F., Sert, E., Avci, E., & Dogantekin, E. (2019). Brain tumor detection based on Convolutional Neural Network with neutrosophic expert maximum fuzzy sure entropy. *Measurement*, 147, 106830.

Razi, F. A., Bustamam, A., & Latifah, A. L. (2023, July). Development of efficient brain tumor classification on MRI image results using EfficientNet. In *2023 International Seminar on Intelligent Technology and Its Applications (ISITIA)* (pp. 575–580). IEEE.

Sangeetha, R., Mohanarathinam, A., Aravindh, G., Jayachitra, S., & Bhuvaneswari, M. (2020, November). Automatic detection of brain tumor using deep learning algorithms. In *2020 4th International Conference on Electronics, Communication and Aerospace Technology (ICECA)* (pp. 1–4). IEEE.

Sasikala, M., & Kumaravel, N. (2008). A wavelet-based optimal texture feature set for classification of brain tumours. *Journal of Medical Engineering & Technology*, 32(3), 198–205.

Seetha, J., & Raja, S. S. (2018). Brain tumor classification using convolutional neural networks. *Biomedical & Pharmacology Journal*, 11(3), 1457.

Chapter 14

IoT-empowered video surveillance: Enhancing security with WiMAX technology

Debashish Tiwary, Avisikta Mandal Das, Namita Shaw, Ayes Chinmay, and Anmol Pattanaik

14.1 INTRODUCTION

Rapid technologies in the field of IoT have made modern systems of security drastically different than before. With IoT-assisted video surveillance, what is achieved now is unrivalled with regard to real-time monitoring, analyzing data, and detecting possible threats. All these integrated cameras, sensors, and computer devices communicate effectively and synchronize their activities over a united network distributed from multiple locations. However, such a system does pose challenges as far as deployment is involved, such as high-bandwidth needs, reliable connection, and scalability in quite diverse environments.

The growing demand for real-time video data transfer and the rising density of IoT devices require strong communication technologies. Here, WiMAX is an emerging option. It is designed for high-speed wireless broadband coverage over large areas, bringing reliability, low latency, and high throughput to a data-intensive application like video surveillance through IoT.

14.1.1 Relevance of IoT and WiMAX

IoT-powered video surveillance systems are complex. They need reliable data transmission for quick responses. IoT devices connect to each other, which allows remote system checks smart video analysis and better security. These systems need a strong communication backbone that can handle lots of video data without slowing down or losing quality.

WiMAX technology meets these needs. It offers wireless broadband over large areas. It supports high data speeds and many connections at once. This solves the main problems of old wireless communication tech in video surveillance. Also, you can set up WiMAX in cities, countryside, or far-off places. This makes it a good fit for IoT-based systems.

DOI: 10.1201/9781003589716-14

14.1.2 Problem statement

IoT-enabled video surveillance systems even though have a lot of demand due to their effectiveness still face certain obstacles.

1. **Bandwidth management**: Due to requirements of video outputs data rate, video footage can be limited because of network overload.
2. **Latency**: For there to be threat detection in real time and timely response to it is in demand, there needs to be minimal amounts of latency involved.
3. **Technical limitations**: With the growth of IoT device performance, consistency is harder and harder to achieve.
4. **Physical barriers**: There are also physical barriers that wireless networks are affected by which include obstructions and interference.
5. **Energy constraints**: A long-term solution to keep IoT devices and gateways functional includes ideal energy solutions.

Resolving these issues calls for a holistic view of the capacity and performance interactions of IoT-based surveillance systems, especially when these systems are the focus of research on WiMAX technology.

14.1.3 Objectives of the study

The aim of this study is to survey the analysis of the capacity and the performance of IoT video surveillance system on WiMAX technology. Particular objectives include:

1. **Throughput and latency evaluation**: Striving to accommodate operational loads with enough input and least response time.
2. **Scalability analysis**: Encompassing the assessment of 'Scaling-Up' the credible supercomputer from a lower configuration.
3. **Quality of service assessment**: Making sure that the aspects of communication real-time and robust are well taken care of.
4. **Bandwidth optimization**: Increasing the utilization of bandwidth by both mobile and computer network.
5. **Environmental factor impact**: The placing conditions, as well as environmental conditions, are analyzed and how they affect the responsive capability of the system are also considered.
6. **Energy Efficiency**: Investigating eco-friendly energy technologies for IoT devices.

14.1.4 Purpose of the research

This study is significant in enhancing the deployment of IoT-based video surveillance technologies. Its goal is to deploy WiMAX technology in order to deal with the existing challenges and recommend improvements for security systems. Therefore, the knowledge will aid in developing more efficient, scalable, and reliable systems, thus facilitating their incorporation into smart city's industrial zones and protection of critical assets.

Furthermore, the research intends to provide useful recommendations to professionals and policy-makers so that these advanced surveillance systems can be implemented. This is in line with the overall vision of making the neighborhoods safer, smarter, and more interconnected.

14.2 LITERATURE SURVEY

IoT-based video surveillance systems use a network of connected cameras and sensors to provide full monitoring solutions. These systems allow for remote and non-stop surveillance, which is key for security in public spaces, private organizations, and homes (Jasim & Atia 2022). The use of edge computing in these systems is important, as it allows data processing to happen closer to where it's collected, which cuts down on delays and how much bandwidth is used. This is essential for real-time analysis, which makes surveillance systems better at responding to unusual events they detect (Liu 2023; HUANG, 2023). The use of edge devices allows for rapid data processing, enabling timely alerts and actions in response to security threats (Liu, 2023).

Moreover, research has led into the development of enhanced algorithms with anomaly detection which is improving the effectiveness of IoT-based video surveillance. The potentiality, for example, for a hybrid combination of CNNs with an ESN is proven in enhancing the security framework for a smart city by detecting anomalies within such large video datasets (Islam et al., 2023). This not only enhances the exactness of detection but also makes surveillance systems more efficient by reducing false alarms and optimizing the usage of resources (Islam et al., 2023).

The ability of WiMAX technology to provide high-speed wireless broadband access over long distances makes it complement IoT video surveillance systems by providing reliable data transmission. This is very helpful in urban environments where the traditional wired connection is not practical. The integration of WiMAX with IoT surveillance systems can allow for smooth and efficient transfer of high-definition video feeds to the processing units, thereby allowing higher quality data to be retrieved during surveillance (Diratie et al., 2021; Chen et al., 2022). Additionally, the integration of WiMAX with edge computing allows the creation of a much stronger architecture that supports the processing and analysis of streams of video in real time. This is critical for an efficient security management system (Ke et al., 2021).

Security and privacy issues are now the ground for major impediments in the implementation of IoT video surveillance. The system must have security in communication and storage since data through the surveillance cameras is sensitive. Literature findings indicate that the data for the activity under surveillance needs to be protected with suitable encryption and access policies. This may ensure lack of access by unauthorized bodies (Caruccio et al., 2020; Khan et al., 2020). Besides, the application of blockchain technology has been suggested to improve the integrity and authenticity of video data so that the footage is tamper-proof (Khan et al., 2020).

Finally, the application of IoT in video surveillance systems is another innovative means of enhancing security management through the use of the WiMAX application. The further development of edge computing, anomaly detection algorithms, and the security mechanisms of transmitting data adds to the design of surveillance solutions that work efficiently and reliably. Future work areas will be ease of addressing security and privacy concerns to keep discovering innovative technologies that can improve the video surveillance systems supported through the Internet of Things.

14.3 METHODOLOGY

This section outlines the systematic approach used in the analysis of the capacity and performance of IoT-based video surveillance systems using WiMAX technology. Architectural design, experimental setup, and the performance metrics used in evaluating the system capabilities under different conditions are shown below.

14.3.1 IoT video surveillance system

Architectural Design IoT architecture for video surveillance system includes certain components which are as follows.

14.3.1.1 IoT-enabled cameras

- For capturing the video feeds, the cameras which have IoT-enabled features are equipped (e.g. ESP 32 Cam module)
- Advanced features such as night vision, motion detection, video compression, etc. are included.

14.3.1.2 IoT gateway

- Annotated as a data fusion box that receives video streams from several cameras.
- Sends video information to the WiMAX network and does some data processing on the video.

14.3.1.3 *WiMax base stations*

- Supports high-speed wireless broadband to meet long-range data transmission requirements.
- Ensuring secure and fast connection between the IoT gateway and a central control system.

14.3.1.4 *Central monitoring system*

- This comprises the integration of on-premises servers with cloud infrastructure which is the practical aspect of video storage and video analyzing system.
- This deploys machine learning-controlled analytics and other functions for analytics purposes such as action analysis as well as object identification.

14.3.1.5 *User interfaces*

- For viewing the live video from the cameras, a user interface is needed.
- The interactive part will be used on phones and web applications.

14.3.2 Experimental setup

14.3.2.1 *Simulated environment*

- The system is designed using network simulation tools such as NS 3 or MATLAB.
- System testing has been carried out with multi-layer approach combining network plus device plus environment aspects.

14.3.2.2 *Test scenarios*

- Scenario 1: A limited number of IoT cameras in a confined area for use throughout a coverage region with a proper design distribution.
- Scenario 2: Having the installation of many cameras in a wide area of city.
- Scenario 3: Distribution of cameras and network loads are medium.
- Scenario 4: A rural installation which experiences harsh environmental settings and placement configuration.

14.3.2.3 *Configurations*

- Allocating bandwidth based on what would impact delay and throughput.
- Use of priority-based packet scheduling for QoS.
- Using environmental parameters like interference, obstruction, and shadowing.

14.3.3 Performance metrics

In this section, three sets of indicators along core aspects of system complying to the pre-defined guidelines are presented which are Discrete Point of Interest Framework, System engineering decomposed framework, and System architecture framework.

14.3.3.1 Throughput

- The throughput is defined as the amount of original signal that can be transferred from the components to the master control unit.
- With dependence on changes in video feeds or network utilization levels.

14.3.3.2 Latency

- The focus is on the time taken for the cameras attached to the monitoring center to transmit video data.
- Respectable for time-sensitive reactions like security breach or retrieving action.

14.3.3.3 Scalability

- Provides the size and conclusion of increasing the IoT devices while still being able to keep functional the system.
- Instituted by step increasing the number of IoT cameras on the network.

14.3.3.4 QoS

- The indicators employed include loss of packets, jitter, and reliability.
- Promises consistency of video images and the system's stability overall.

14.3.3.5 Bandwidth utilization

- Estimates the effectiveness of the bandwidth allocated to video signal transmission.
- Considers the impacts of loss-less compression techniques and adaptation of bit rate control.

14.3.3.6 Energy efficiency

- Evaluates the power consumption by the IoT cameras and IoT camera's gateways.
- Offers recommendations for improving energy efficiency while complying with the required performance levels.

14.3.4 Analytical techniques

14.3.4.1 Statistical analysis

- Simulation produced data is subjected to statistical methods in order to establish patterns and relationships.

14.3.4.2 Comparative analysis

- By using the technologies, the performance in comparison of the new system against Wi-Fi and the LTE is tested.

14.3.4.3 Optimization models

- Processes are established with the aim of increasing the efficiency of allocation of resources such as bandwidth and electricity.

14.3.4.4 Sensitivity analysis

- Some factors such as distance to the transmitter, interference, and the number of devices in an area have an effect on performance of the system.

14.3.5 Implementation framework

14.3.5.1 Prototype development

- Systematic check of the results of the simulated experiments as illustrated by the introduction of a prototype of the system.
- Secured IoT equipment, gateways, and a WiMAX network are used in controlled environments.

14.3.5.2 Conducting tests in the field

- In order to understand its feasibility and effectiveness, the system is deployed in a real-world situation.

14.3.5.3 Receiving feedback and reiterating

- Feedback concerning performance is registered and is generally used to improve system design and its overall effectiveness.

14.3.5.4 Ethical and security considerations

- **Privacy:** Guarantees adherence to data protection requirements by the anonymization of sensitive information.

- **Security**: Employs encryption and secure communication techniques to safeguard video data.
- **Sustainability**: Assesses ecological consequences and advocates for energy-efficient methodologies in system implementation.

14.4 RESULTS

This section addresses the analysis and experimentation conducted on the WiMAX-based IoT video surveillance system about the obtained results, categorized under key performance metrics such as throughput, latency, scalability, QoS, bandwidth usage, impacts on the environment, and energy efficiency.

14.4.1 Throughput analysis findings

- The system delivered on average **25Mbps** in low-density deployments and thus can support real-time streaming of HD video (Mishra et al., 2019).
- It reduced throughput to **15Mbps** in high-density deployments because of increased network traffic.
- The video compression techniques using H.265 increased throughputs by about 20% while reducing the workload on the network.

14.4.1.1 Implications

Efficient data encoding is essential for sustaining high throughput in IoT-based video surveillance systems, particularly in environments with several devices.

14.4.2 Latency characteristics

14.4.2.1 Findings

- WiMAX technology had less than **50 ms** of latency, so quite useful when monitoring video real time.
- However, environmental factors and data transmission distance did impact the speed, with latency rising to **120 ms**.
- The use of edge computing resulted in **30% increase** in reaction times, as the requirement to transfer data to a central database was eliminated; instead, all necessary information was processed locally.

14.4.2.2 Implications

For applications where there are strict time requirements like intruder alert, edge devices must be close to the IoT cameras.

14.4.3 Scalability assessment findings

- The results of the test showed stable performance ability of the system that connected 200 IoT devices to the WiMAX base station.
- Once this limit has phase congestion, there is an increase in packet loss and a decrease in throughput.
- The resources of the network were well distributed, enhancing the performance of the system due to dynamic bandwidth allocation techniques.

14.4.3.1 Implications

The system is suitable for deployments of an average size; however, large deployments require better management of resources.

14.4.4 QoS

14.4.4.1 Findings

- Sangita karela testing under two conditions managed to provide an assessment of 99.2% packet delivery reliability.
- Michael W. Haigh et al. agreed that J. Zhao could not exceed values of 10ms yielding smooth video playback performance.
- According to prior research, QoS was compromised and congestion reached 5% and even more after the highest peak was reached.

14.4.4.2 QoE and fundamental insight

By prioritizing the backbone cabling as the dominant requirement and implementing adaptive QoS approaches, consistency is assured despite the ceaseless fluctuations in the demand on the network.

14.4.5 Bandwidth utilization findings

- Bandwidth utilization averaged 75% in the case of low-density scenarios and reached 90% in case of high-density deployments.
- Using a streaming adaptable bitrate resulted in a decrease of 25% in bandwidths without affecting the video quality (R. Mahanty et al., 2019).
- Use of effective video compression techniques as well as dynamic allocation of resources contributed to higher levels of effective bandwidth.

14.4.5.1 Implications

These measures for optimization of bandwidth resources allocation would be quite critical in ensuring that the system remains efficient especially during high traffic situations.

14.4.6 Environmental impact findings

- Physical complexities like buildings and trees reduced the strength of the signal by as high as 30% and as a result impacted on both throughput and latency levels negatively.
- The distance between neighboring WLANs caused the reliability of data transfer to fall by 15% (Rai P. et al., 2020).
- The sight positioning of WiMAX base stations virtually eliminated these problems.

14.4.6.1 Implications

For the viable performance in extreme environments, the selection of the ideal placement of base stations and interference suppression measures are fundamental.

14.4.7 Energy efficiency findings

- Average power consumption of IoT cameras and IoT gateway was **10 W** and **15 W**, respectively (Srivastava A. et al., 2020).
- Use of low-power devices and sleeping modes reduced energy consumption by 20% and 30%, respectively.
- Solar-powered gateways appear to be suitable for off-grid pilot projects, providing energy alternatives that are renewable.

14.4.7.1 Implications

Energy conserving designs as well as sources of energy such as solar and wind in addition to other renewable energies are very much necessary in conserving economic sustainability of operations in the future.

14.4.8 Differences of other technologies

14.4.8.1 WiMAX with Wi-Fi

- WiMAX has larger coverage area compared to Wi-Fi coverage area and hence was able to connect up to **50 km** while Wi-Fi would connect up to **300 m** (Chinmay and Pati, 2024a; Chinmay and Pati, 2024b; Chinmay and Pati, 2024c).
- In high-density deployments, WiMAX is able to show 35% increase in the throughput level.

14.4.8.2 WiMAX to LTE

- LTE has relatively low latency when data transfer was done, but it was more costly in large deployments such as infrastructures because they

required more components (Chinmay and Pati, 2024d; Chinmay and Pati, 2024e; Chinmay and Pati, 2024f).
- WiMAX performance is retained but at a cheaper price, which made it effective for city and country use.

14.4.9 Overall system performance

The integration of IoT and WiMAX technology provided dependable and scalable video surveillance solutions. The system exhibited robust performance in situations necessitating real-time monitoring with minimal latency. Also, the system supports high-density devices with optimized demands like optimal bandwidth management and low energy utilization.

Table 14.1 represents the summary of the results used in this chapter.

14.5 CONCLUSIONS AND FUTURE SCOPE

In video surveillance systems, IoT technology has been integrated with WiMAX which has proven a lot against modern security challenges. The capacity and performance of IoT-based video surveillance systems were evaluated in this study, particularly concerning important metrics such as throughput, latency, scalability, QoS, bandwidth utilization, environmental

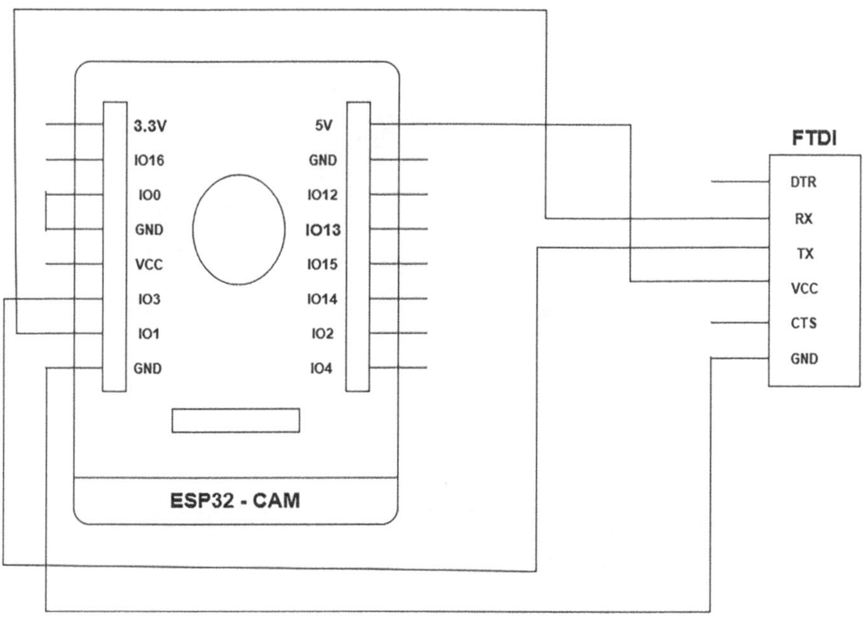

Figure 14.1 Schematic layout of connections between ESP32-CAM and FTDI.

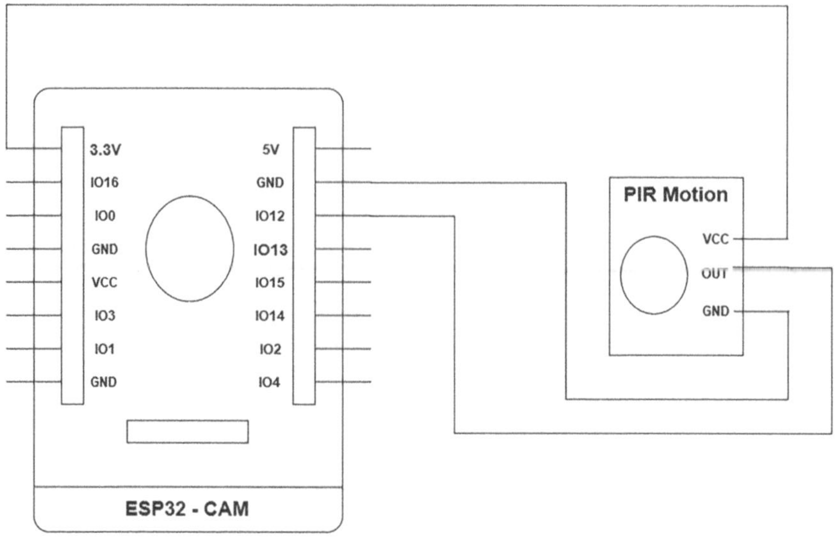

Figure 14.2 Schematic layout of connections between ESP32-CAM and PIR sensor.

Figure 14.3 Wired connection between FTDI232 and ESP32-CAM.

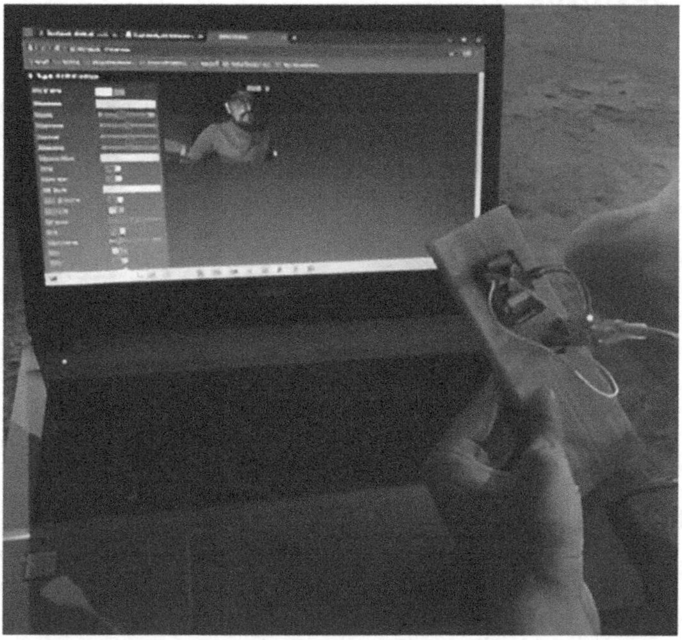

Figure 14.4 Live streaming of ESP32-CAM.

Table 14.1 Summary of results

Metric	*Key Finding*	*Optimization*
Throughput	25 Mbps (low density), 15 Mbps (high density)	Video compression (H.265)
Latency	<50 ms (ideal), 120 ms (interference)	Edge computing, strategic base station placement
Scalability	Up to 200 devices per base station	Dynamic bandwidth allocation
QoS	99.2% reliability, <10 ms jitter	Adaptive QoS policies
Bandwidth Utilization	75–90% utilization	Adaptive bitrate streaming
Environmental Impact	30% signal reduction (obstructions)	Line-of-sight deployment
Energy Efficiency	Reduced power by 20–30%	Sleep modes, solar-powered gateways

impacts, and energy efficiency. These results have proven that the WiMAX technology could also become an alternative backbone for wireless surveillance IoT systems where it's necessary to cover large areas in real time, and where devices' density is high.

KEY TAKEAWAYS

1. **Performance Capabilities:**
   - WiMAX has high throughput and low latency which is necessary for live video streaming in crowded IoT spaces.
   - The system was scalable in nature, where a single base station could support 200 devices while experiencing little to no scaling down of device performance.
2. **Efficiency Enhancements:**
   - Video compression techniques together with adaptive bitrate streaming and dynamic resource allocation improved the efficiency of bandwidth usage and lessened the energy spent.
   - Latency is reduced by edge computing technology and improved the real-time engagement which makes the system a good candidate for urgent security functions.
3. **Climatic Adaptability:**
   - Some factors like interference from signage and environmental constraints did limit performance at lower levels, the placement of WiMAX base station and deploying interference preventing strategies helped in diminishing the level of poor performance.
4. **Comparison with Alternative Technologies:**
   - Global positioning technology developed at the university can sensor be drawn across several miles instead of several hundred feet beating Wi-Fi, whereas anywhere WIMAX is available will beat LTE which cost wise is important for video surveillance systems to use rural and urban scenery.

14.5.1 Contributions to security infrastructure

The study demonstrates how many IoT-enabled video surveillance systems could be possible within the WiMAX network (i.e., capacity analysis of WiMAX considering IoT-enabled video surveillance systems). IoT-enabled video surveillance systems are crucial for transferring large amounts of data with minimal latency and for facilitating numerous installations, making them significant components in enhancing security across various domains.

These areas include:

- **Smart Cities:** Allowing the ability to monitor most parts of the city, control movement of traffic, and respond to emergency situations.
- **Critical Infrastructure:** Providing effective means of surveillance of energy facilities, transport hubs, and industrial sites.
- **Remoter Areas:** Contributing to the deployment of reliable and quality video monitoring in rural and underdeveloped areas irritated with less infrastructure to support economical solution.

14.5.2 Limitations

Despite the advantages offered by the electronics developed during the study, certain disadvantages were also acknowledged by the report:

- **Signal Attenuation**: Structures and interference are major issues affecting the performance of wireless devices operating in high-frequency bands.
- **Power Dissipation**: The amount of energy consumed by electrical equipment is manageable but to develop more extensive and practical ideas, greater energy boosting measures would be required.
- **Scalability Criteria**: When the number of devices as a measure of density is increased above a certain threshold, optimization techniques of some sophistication become necessary as performance starts to degrade.

14.5.3 Future directions

Hence, many questions arise from these results, which makes it an interesting area for future research.

1. **5G**: The use of 5G technologies would be beneficial toward performance – latency and number of devices supported.
2. **AI-Supported Applications**: Utilizing AI and Machine Learning for Enhanced Video Analytics, Anomaly detection services, and Comprehensive Threat Analysis (Mohapatra N. et al., 2020).
3. **Edge and Fog Computing**: Greater use of edge and fog computing would reduce central energy consuming server dependence and further lower latency.
4. **Sustainability Measurements**: Looking into IoT systems powered by renewable energy resources like photovoltaic cells to achieve the most environmentally friendly and energy-efficient system operation.
5. **Advanced Security Protocols**: Strong systems of encryption and authentication that will protect video information and prevent unauthorized access.

REFERENCES

Caruccio, L., Piazza, O., Polese, G., & Tortora, G. (2020). Secure IoT analytics for fast deterioration detection in emergency rooms. *IEEE Access*, 8, 215343–215354. 10.1109/access.2020.3040914

Chen, Y., Lin, Y., Hu, Y., Hsia, C., Lian, Y., & Jhong, S. (2022). Distributed real-time object detection based on edge-cloud collaboration for smart video surveillance applications. *IEEE Access*, 10, 93745–93759. 10.1109/access.2022.3203053

Diratie, E., Sharma, D., & Agha, K. (2021). Energy aware and quality of service routing mechanism for hybrid internet of things network. *Computers*, 10(8), 93. 10.3390/computers10080093

Huang, J. (2023). Investigating of deep learning-based approaches for anomaly detection in IoT surveillance systems. *International Journal of Advanced Computer Science and Applications*, 14(12). 10.14569/ijacsa.2023.0141279

Islam, M., Dukyil, A., Alyahya, S., & Habib, S. (2023). An IoT enable anomaly detection system for smart city surveillance. *Sensors*, 23(4), 2358. 10.3390/s23042358

Jasim, M. and Atia, T. (2022). An IoT-fuzzy based password checker system for wireless video surveillance system. *Bulletin of Electrical Engineering and Informatics*, 11(6), 3441–3449. 10.11591/eei.v11i6.4375

Chinmay, A. and Pati, H. K. (2024a). VoWiFi cell capacity evaluation using WiFi 7 considering VBR traffic. *IETE Journal of Research, Taylor and Francis.* 10.1080/03772063.2023.2291793

Chinmay, A. and Pati, H. K. (2024b). Voice over WiFi cell capacity enhancement using A-MPDU frame aggregation in WLAN standard. *Peer-to-Peer Networking and Applications*, Springer. 10.1007/s12083-024-01628-8

Chinmay, A. and Pati, H. K. (2024c). Enhancement of VoWiFi cell capacity using A-MPDU frame aggregation technique in WiFi 6 considering VBR traffic. *International Journal of Communication Systems*. 10.1002/dac.5782

Chinmay, A. and Pati, H. K. (2024d), VoWiFi cell capacity using A-MPDU frame aggregation in sixth generation WLAN standard. *ETRI Journal, Wiley.* 10.4218/etrij.2023-0333

Chinmay, A. and Pati, H. K. (2024e). Capacity analysis of a WLAN cell using VoWiFi service for CBR Traffic. *Wireless Personal Communications*. 10.1007/s11277-024-11478-5

Chinmay, A. and Pati, H. K. (2024f), "*VoWiFi Cell Capacity Improvement Using A-MPDU Frame Aggregation Technique for VBR Traffic*", Concurrency and Computation: Practice and Experience, Wiley. 10.1002/cpe.8247

Ke, R., Zhuang, Y., Pu, Z., and Wang, Y. (2021). A smart, efficient, and reliable parking surveillance system with edge artificial intelligence on IoT devices. *IEEE Transactions on Intelligent Transportation Systems*, 22(8), 4962–4974. 10.1109/tits.2020.2984197

Mohapatra, N., Shreya, K., and Chinmay, A. (2020). Optimization of the Random Forest Algorithm. In: Borah S., Emilia Balas V., Polkowski Z. (eds) *Advances in Data Science and Management. Lecture Notes on Data Engineering and Communications Technologies*, vol 37. Springer, Singapore. 10.1007/978-981-15-0978-0_19

Srivastava, A., Khare, A., Satapathy, P., and Chinmay, A. (2020). Investigating Various Cryptographic Techniques Used in Cloud Computing. In: Borah S., Emilia Balas V., Polkowski Z. (eds) *Advances in Data Science and Management. Lecture Notes on Data Engineering and Communications Technologies*, vol 37. Springer, Singapore. 10.1007/978-981-15-0978-0_26

Rai, P., Prasad, A., Reddy, S.M., and Chinmay, A. (2020) Evolution of optical storage in computer memory. In: Borah S., Emilia Balas V., Polkowski Z. (eds) *Advances in Data Science and Management. Lecture Notes on Data Engineering and Communications Technologies*, vol 37. Springer, Singapore. 10.1007/978-981-15-0978-0_48

Mahanty, R., Mahapatra, S., Nayak, A., and Chinmay, A. (2019). Comparative Study of Various Image Captioning Models. *International Conference on Recent*

Advances in Energy-efficient Computing and Communication (ICRAECC), pp. 1–5. 10.1109/ICRAECC43874.2019.8995014

Mishra, S., Sethi, N., and Chinmay, A. (2019). Various Data Skewness Methods in the Hadoop Environment. *International Conference on Recent Advances in Energy-efficient Computing and Communication (ICRAECC)*, pp. 1–4. 10.1109/ICRAECC43874.2019.8994979

Khan, P., Byun, Y., and Park, N. (2020). A data verification system for CCTV surveillance cameras using blockchain technology in smart cities. *Electronics*, 9(3), 484. 10.3390/electronics9030484

Liu, H. (2023). An edge computing-based handgun and knife detection method in IoT video surveillance systems. *International Journal of Advanced Computer Science and Applications*, 14(11). 10.14569/ijacsa.2023.0141117

Chapter 15

Automatic gender detection: An AI perspective

Binodini Tripathy, Niva Das, Pratyush Ranjan Nayak, Unnayan Raj, and Saurav Datta

15.1 INTRODUCTION

The technology for biometric identification has increased significantly in the last few years. In identity recognition, the use of finger print and facial image scanning has substantially increased. Though these forms of identity recognition have been quite effective, flaws in these propel for designing different forms of biometric identification. Voice-based identity recognition or biometric identification is a more popular form of identity recognition now a days. Major factors behind are accuracy, reliability, speed, convenience, and privacy. The accuracy of the voice-based identification is by far the best (up to 98%) compared to facial and finger print scanning. The reliability of these systems has also shown minimum false acceptance rate and minimum false rejection rate. Compared to other modalities, voice-based system also operates very fast. Voice-based systems are also a very convenient form of identity recognition as these do not require any complex training procedure or software as is required for face detection and finger print identification. Privacy is another major concern which makes the voice-based identity recognition system most user friendly. These systems do not produce any recording or image that could be used for doing fraud. Companies like Apple, Microsoft, Google, Amazon, and Samsung have developed voice-based biometric systems which are quite popular and efficient. These promising features of voice-based systems make it popular among the scientific and AI research community for several other applications like automatic gender detection, speaker recognition, speech recognition, etc. [1]. This chapter presents the problem of automatic gender detection from an AI perspective based on voice samples.

Automated detection of human gender plays a crucial role in many frameworks that deal with processing of real-time human-related data as in human-machine interaction. Due to the extensive use of social media platforms, automated gender detection has become relevant and essential to a variety of applications. Some of the applications worth mentioning are gender categorization-based sorting of telephone calls in call centers, video

 DOI: 10.1201/9781003589716-15

categorization and tagging, virtual assistant systems, security systems, criminal discrimination systems, etc.

The problem of gender detection deals with identifying the gender (male/female) of the human being from available person-related resources. One of the prominent sources from which gender of a person can be identified is the sound characteristics. Gender information also influences processing of speech samples. Nowadays automated computer systems are being deployed in areas like healthcare, education, law for better management and control. Healthcare system comprising gender detection subsystems can offer more options and provide more accurate information.

Gender detection from speech can be viewed as a binary classification task where relevant features are extracted from speech samples of the speaker followed by employing an appropriate ML- or DL-based classifier to identify the speaker as male or female. Many studies have been carried out on the subject focusing on finding optimum features and designing suitable ML and DL classifier models. Most commonly reported speech features for gender detection task are pitch, formant frequency, MFCC, MEL, chroma, Linear Prediction Cepstral Coefficients (LPCC), etc. [2] ML-based approaches like SVM, K-NN, Gaussian Mixture Model (GMM), Random Forest (RF) and Artificial Neural Nets have also been used for classification of gender [3]. Deep learning-based classifier models are becoming more popular due to their superior recognition ability, resulting in higher accuracy [4]. Among DL-based models, CNN, RNN, and LSTM have shown better performance.

In this chapter, our prime objective is to explore only single but powerful features like pitch and MFCC and test their recognition ability on classical ML and DL classifier models. Two datasets namely RAVDESS and Mozilla Common Voice have been considered for feature extraction, training, and testing the ML and DL models. The remaining part of the chapter is organized into four sections. Literature survey is presented in Section 15.2. Section 15.3 briefly describes the datasets, steps of preprocessing and feature extraction, ML and DL classifier models used. Section 15.4 elaborates on the results of the models used. Section 15.5 concludes the work along with the scope for any possible extension.

15.2 LITERATURE REVIEW

Several time domain and frequency domain features have been reported in the literature for the task of voice-based gender detection. Shabbir et al. [5] report a comparative analysis of three features based on age and gender detection models. The features considered are pitch, MFCC, and Spectral Dynamic Coefficients (SDC). SDC-based SVM classifier with radial basis function (RBF) kernel achieves 97.5% accuracy, whereas the same classifier based on pitch features achieves 97.3% accuracy. In ref. [6], a combination of prosodic features such as energy, pitch, formant frequency along with MFCC have been used to detect gender. The use of a greater number of uncorrelated features helps in enhancing the detection accuracy. In ref. [7], the proposed system relies on MFCC as well as its first- and second-order derivatives and achieves 98.4% accuracy of detection with the help of an ensemble of SVM and Neural Net Classifier. Mavaddati [8] present a comparative analysis of DL models, i.e., CNN, RNN, ResNet-34 based on recorded audio samples for the dual task of detecting gender and age. In ref. [9], ML classifier models, i.e., SVM, k-NN, ANN, GMM, and RF, have performed very well in classifying gender from speech samples. In ref. [4], the authors have used 20 acoustic features extracted from voice samples to detect gender using ANN classifier. Pretrained DNN models like ResNet-50 and ResNet-34 have performed well on spectrogram data along with hand-crafted features [10]. In ref. [11], the authors have proposed a double layer LSTM and claim to have achieved 98.4% average accuracy on a dataset consisting of 20 features against each voice sample.

15.3 MATERIALS AND METHODS

15.3.1 Materials

15.3.1.1 Datasets

RAVDESS is a multimodal database containing emotional speech in English in North American accent. 12 male and 12 female professional actors have given their voice to build it. Since it is a gender-based dataset, we have used only the speech samples for our work. The speech samples are available as .wav files. The second dataset used in this work is Mozilla Common Voice dataset which contains recorded audio of volunteers reading texts from publicly available sources supporting over 93 languages. All the recorded audio data are not labeled properly. The dataset comes with CSV files containing information on gender along with others. Among the properly labeled audio samples, a total of 70K samples comprising 35K female and 35K male data are selected and used in this work.

15.3.2 Methods

Figure 15.1(a) and (b) display the block diagram representation of the scheme proposed.

15.3.2.1 Preprocessing

Before extraction of potential features, the speech signals are preprocessed. Preprocessing involves steps like noise removal, silence removal, normalization, windowing, etc.

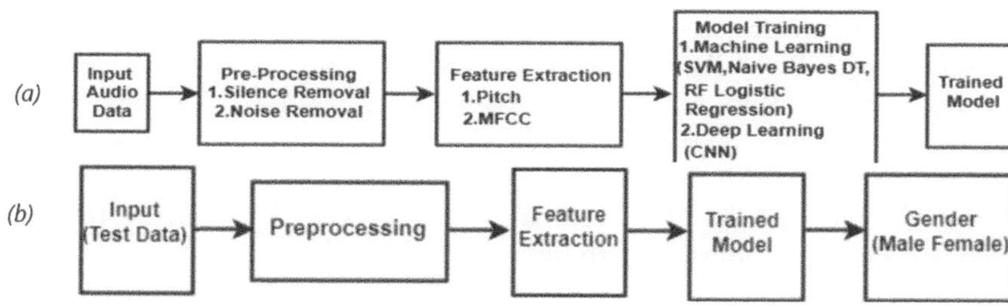

Figure 15.1 (a) Block diagram representation of the training model of gender-detection system. (b) Block diagram representation of the testing model of gender-detection system.

15.3.2.2 Noise removal

Noise removal in audio processing aims to enhance the quality and intelligibility of audio by minimizing unwanted background sounds. Common techniques include spectral subtraction, where the noise spectrum is estimated and subtracted from the audio signal and Wiener filtering, which adjusts the signal based on a statistical model of the noise. To remove noise Butterworth filter, spectral gating technique, SNR filter, and DeepFilterNet have been tested for our work. However, due to faster and better noise reduction capability, DeepFilterNet and SNR filter are found to be more suitable for the task and hence have been used [12].

For comparing different noise reduction techniques, Signal-to-Noise Ratio (SNR) has been used as an objective measure of noise present in the signal. The SNR can be calculated as follows.

- Power of signal

$$P_{signal} = \frac{1}{N}\sum_{i=1}^{N} x[i]^2$$

Where $x[i]$ are the samples of the signal and N is the total number of samples.

- Power of noise

$$P_{noise} = \frac{1}{M}\sum_{j=1}^{M} n[j]^2$$

Where $n[j]$ are the samples of the signal and M is the total number of samples.

- SNR:

$$SNR(dB) = 10 \times \log_{10}\left(\frac{P_{signal}}{P_{noise}}\right)$$

The following noise removal techniques have been tested.

Butterworth Band Pass Filter. A Butterworth band pass filter is designed to pass frequencies within a specified range (the pass band) while attenuating frequencies outside that range (the stop bands). The Butterworth filter is known for its smooth frequency response, characterized by having no ripples in the pass band and a maximally flat response. The transfer function H(s) for a low-pass Butterworth filter is:

$$H(s) = \frac{1}{\sqrt{1 + \left(\frac{s}{\omega_c}\right)^{2n}}}$$

where ω_c is the cutoff frequency, s is the complex frequency variable, and 'n' is the filter order. For a band pass filter, the design involves a combination of low-pass and high-pass filters. The band pass filter allows frequencies between ω_{low} and ω_{high} to pass presented by the transfer function:

$$H(s) = \frac{1}{\sqrt{1 + \left(\frac{\omega_{low}}{s}\right)^{2n}} \times \sqrt{1 + \left(\frac{s}{\omega_{high}}\right)^{2n}}}$$

The Butterworth band pass filter ensures a flat response in the pass band, making it suitable for applications requiring minimal signal distortion.

Spectral Gating. Spectral gating is a noise reduction technique that operates in the frequency domain. The audio signal is first transformed using the Short-Time Fourier Transform (STFT) to obtain its time frequency representation.

$$X(k,m) = STFT(x(t)) \mid$$

where X(k,m) represents the frequency bin k at time frame m. A threshold is applied to each frequency bin and components below this threshold are attenuated or set to zero.

$$\begin{cases} 1 \; if \; |X(k,m)| \geq threshold \\ 0 if \; |X(k,m)| < threshold \end{cases}$$

The filtered signal is obtained by:

$$Y(k,m) = X(k,m) \cdot G(k,m)$$

The inverse STFT is then applied to transform the modified spectral components back to the time domain, resulting in a noise-reduced signal.

SNR Filter. An SNR filter enhances the quality of a signal by emphasizing components with a higher SNR. SNR is a measure of the relative power of the signal to the noise and is expressed in decibels (dB).

$$SNR(dB) = 10 \times log_{10}\left(\frac{P_{signal}}{P_{noise}}\right)$$

Where P_{signal} is the power of the signal and P_{noise} is the power of the noise. In practice, the SNR filter applies a weighting function based on the calculated SNR, boosting components with higher SNR and attenuating those with lower SNR:

$$Y(t) = x(t) \cdot W(t)$$

where $W(t)$ is the weighting function derived from the SNR. This process improves the overall clarity and intelligibility of the signal by reducing the impact of noise.

DeepFilterNet: DeepFilterNet is a neural network-based system designed for real-time speech enhancement. It employs a deep learning approach to filter out background noise and improve the quality of speech signals. The model architecture typically includes convolutional layers and recurrent units to capture both spatial and temporal features of the audio signal. By training on noisy speech datasets, DeepFilterNet learns to distinguish between speech and noise, enhancing the former while suppressing the latter. The network processes input frames and applies learned filters, outputting cleaner speech. This technology is particularly useful in applications like telecommunication, hearing aids, and voice-activated systems.

15.3.2.3 Silence removal

Silence removal operation involves retaining non-silent intervals achieved by setting a threshold for silence in terms of decibel relative to the peak amplitude of the voice sample. This operation ensures that only relevant audio content is retained for feature extraction. Silence removal in audio processing enhances clarity and reduces file size. The LibROSA library provides the trim function to achieve this. Librosa effects trimfunction detects non-silent intervals in an audio signal based on amplitude thresholds and returns the trimmed audio along with the indices of the non-silent regions. It can be fine-tuned using parameters like top db, which sets the threshold for silence in decibels relative to the peak. This function is useful for preprocessing in speech and music analysis, ensuring that only relevant audio content is retained for further processing or analysis [13].

Mathematically, it operates by analyzing the amplitude envelope of the audio signal y(t). It sets a threshold θ and removes segments where the amplitude $|y(t)| < \theta$. After testing, a threshold value of 20 dB was taken. A silence-removed audio sample extracted from a raw audio sample is presented in Figure 15.2.

15.3.2.4 Feature extraction

The cleaned signal after silence removal is subjected to processes like framing, pre-emphasis, and windowing. Then pitch and MFCCs are extracted from each frame.

Pitch Extraction. To determine pitch, autocorrelation method has been used. The autocorrelation of speech sample is first calculated at various time lags and then their cumulative average is determined. Minimizing the argument-based autocorrelation function divided by cumulative average gives the pitch as the argument. For each speech frame, pitch is calculated following the same set of steps as above. The pitch vector is fed to the ML classifier for discrimination between male and female.

MFCC Extraction. Mel-Frequency Cepstral Coefficients (MFCCs) are widely used in speech and audio processing to represent the short-term power spectrum of a sound. They mimic the human auditory system's response more closely than raw spectral representations by combining Mel-frequency scaling and discrete cosine transform (DCT). MFCCs capture spectral characteristics, emphasizing perceptually important features and reducing dimensionality. Mel-Frequency Cepstral Coefficients (MFCCs) are computed using a number of steps involving techniques to process signals. Here's a mathematical explanation of the steps involved.

- **Framing the Signal**: The audio signal x(t) is divided into short overlapping frames. Each frame typically lasts around 20–40 milliseconds.

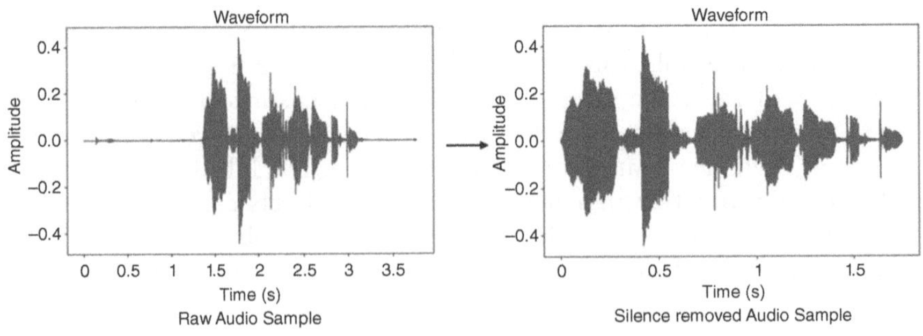

Figure 15.2 Raw audio and silence-removed audio samples.

- **Pre-emphasis:** It is a technique in which a high pass filter is used to eliminate low-frequency noise resulting in a higher SNR.

$$x_{\text{preemp}}(t) = x(t) - \alpha x(t-1)$$

where α is a predefined constant (often 0.97).

- **Windowing:** Each frame is multiplied by a window function (such as Hamming window) to reduce spectral leakage:

$$x_{\text{windowed}}(n) = x_{\text{preemp}}(n) \cdot w(n)$$

where $w(n)$ is the window function.

- **Computation of Power Spectrum:** The Discrete Fourier Transform (DFT) is applied to each frame to obtain its frequency components:

$$X(k) = DFT\{x_{\text{windowed}}(n)\}$$

The power spectrum $P(k)$ is calculated as $|X(k)|^2$.

- **Applying Mel Filter bank:** Mel-scale filters are applied to the power spectrum to convert the frequency scale from Hertz (Hz) to the Mel scale, which better reflects human auditory perception. The Mel filter bank $H_m(k)$ is applied.

$$S_m = \log\left(\sum_{k=0}^{N-1} P(k) H_m(k)\right)$$

where S_m is the Mel-scale log energy.

- **Applying Discrete Cosine Transform (DCT):** Finally, a DCT is applied to decorrelate the Mel-frequency cepstral coefficients:

$$C_n = \sum_{m=0}^{M-1} \log(S_m) \cos\left[\frac{\pi n(m-0.5)}{M}\right]$$

where C_n are the MFCCs and M is the number of Mel bands.

MFCCs capture the essential spectral features of the audio signal, emphasizing perceptually important components while reducing dimensionality, making them suitable for tasks like speech recognition and audio classification. Librosa library and its 'librosa.feature.mfcc' function have been used to extract the MFCCs from each audio frame [14]. After the extraction, each MFCC value is then normalized using the following equation ensuring that each coefficient has zero mean and unit variance across the dataset, which can improve the performance of machine learning models that rely on these features.

$$\text{Normalized MFCC} = \frac{\text{MFCC} - \bar{x}(\text{MFCC})}{\sigma(\text{MFCC})}$$

- MFCC is the original Mel-Frequency Cepstral Coefficient.
- $\bar{x}$ (MFCC) is the mean of the Mel-Frequency Cepstral Coefficients.
- σ(MFCC) is the standard deviation of the Mel-Frequency Cepstral Coefficients.

The normalized MFCC data along with the raw MFCC data are shown in Figure 15.3.

15.3.3 Machine learning models and deep learning models

15.3.3.1 Machine learning models

Machine learning models considered for the two-class gender classification task are SVM [14], DT [15], Logistic Regression [16], Naïve Baye's [17], Ensemble Bagged Tree, and RF [18].

Support Vector Machines (SVMs). Support Vector Machines (SVMs) are powerful classifiers capable of handling non-linear decision boundaries through the use of kernel functions. By mapping input data into a higher-dimensional space, SVMs find optimal hyperplanes that maximize the margin between different classes. This allows SVMs to effectively classify complex patterns that aren't linearly separable in their original feature space. Common kernels include Polynomial, Gaussian Radial Basis Function (RBF), and sigmoid functions, each suited for different types of non-linear reparability in data [14].

Logistic regression. Logistic regression is a fundamental statistical model used for binary classification tasks. It predicts the probability of a binary

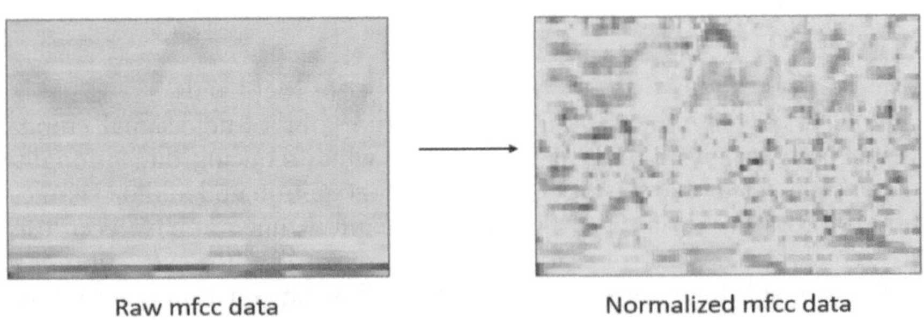

Figure 15.3 Extracted normalized MFCC data from raw MFCC data.

outcome based on input features. It employs the logistic function (sigmoid) to map predictions to probabilities between 0 and 1, allowing for intuitive interpretation. The model estimates coefficients using maximum likelihood estimation, making it computationally efficient even with large datasets. Despite its simplicity, logistic regression assumes a linear relationship between features and the log-odds of the outcome, which can limit its flexibility in complex classification scenarios [16].

***Naïve Byes*:** Naïve Byes is a simple yet powerful probabilistic classifier based on Bayes' theorem with strong independence assumptions between features. It assumes all features are conditionally independent given the class label, simplifying computations and making it efficient for large datasets. Despite its simplicity, Naive Bayes often performs well in text classification and spam filtering tasks, where feature independence approximations hold reasonably well. However, it may struggle with correlated features that violate its independence assumption, potentially affecting classification accuracy [17].

Decision Tree (DT). A decision tree is a versatile ML algorithm used for both classification and regression tasks. It works by recursively partitioning the data into subsets based on the most significant attribute at each node, optimizing toward purity or minimizing error. Each internal node represents a decision based on a feature, while leaf nodes represent the outcome. Decision trees are interpretable, handle both numerical and categorical data, and can model complex relationships with high accuracy, though they're prone to overfitting without regularization [15].

Random Forest (RF). Random Forest is a type of learning based on ensembling approach. Multiple decision trees combine to form RF resulting in improved robustness and accuracy of prediction. A random subset of the data and features is chosen to train each tree in the forest, reducing overfitting and capturing diverse patterns in the data. During prediction, results from all trees are aggregated to make final predictions, providing robustness against noise and outliers. Random Forests are widely used in classification and regression tasks due to their simplicity, scalability, and effectiveness in handling high-dimensional data [18].

15.3.3.2 Deep learning models

The DL-based classifier used for comparison purpose is Convolutional Neural Net [19]. The models have been trained and tested with pitch or MFCC features as input. The models selected have proven performance records in terms of accuracy and prediction time for a wide range of classification and regression tasks.

Convolutional Neural Network. Convolutional Neural Networks (CNNs) are deep learning models designed for processing structured grid-like data, such as images. They consist of convolutional layers that automatically learn spatial hierarchies of features through filters or kernels. These layers are followed by pooling layers to reduce dimensionality and improve computational efficiency. Finally, fully connected layers integrate high-level features for classification or regression tasks. CNNs excel in image recognition, object detection, and medical image analysis due to their ability to capture spatial dependencies and invariant features.

15.3.4 Performance metrics

The performance indices used for evaluating the performance of various ML and DL models are accuracy, precision, recall, F1 Score and average prediction time [10].

$$\text{Accuracy} = \frac{P}{Q}$$

$$\text{Recall} = \frac{A}{B}$$

$$\text{Precision} = \frac{A}{C}$$

$$\text{F1 Score} = \frac{2 \times \text{Precision} \times \text{Recall}}{\text{Precision} + \text{Recall}}$$

where P is the sum of T_P and T_N, Q is the sum of T_P, T_N, F_P and F_N, A is the T_P,B is the sum of T_P and F_N, and C is the sum of T_P and F_P. Average prediction time is the average of the prediction time required to predict the gender from the test samples. In a binary classification task, usually 0 is regarded as negative and 1 is regarded as positive. In our gender classification task, we have considered gender 'male' as positive and the gender 'female' as negative. T_P refers to True Positive, T_N refers to True Negative, F_P refers to False Positive, and F_N refers to False Negative.T_P is the condition when actual data is positive and prediction by the model is positive.T_N is the condition when actual data is negative and prediction by the model is negative. F_P is the condition when actual data is negative, whereas prediction by the model is positive. F_N is the condition when actual data is positive, whereas prediction by the model is negative.

15.4 RESULTS AND DISCUSSION

Pitch being a potential voice-based feature, we started out with pitch vector as input to SVM (Linear), Logistic Regression, DT, and Ensemble Bagged Tree models for training. Fivefold cross-validation is used for training and testing the classifiers on RAVDESS. The implementation is carried out on an Intel-i5 processor with 8 GB RAM in MATLAB R2021a environment, and the results are presented in Table 15.1. Going through the results gives an impression that using pitch as the only feature, maximum accuracy will remain in the range of 80% to 87%.

With a view to improving the performance of the model, we experimented with MFCC features as input to the same ML models on the same RAVDESS dataset, and the results are displayed in Table 15.1. The results ranging from 91% to 99% show tremendous improvement over the previous one, where pitch is considered as the feature.

These results encouraged us to continue our experiment on the two-class classification task with MFCC features. Hence, our next set of results has been obtained on the Mozilla Common Voice dataset. The ML and DL models trained and tested on this dataset are SVM, Logistic Regression, Naive Bayes, DT, Random Forest, and CNN. Since the dataset has much more samples compared to RAVDESS, the training of the models has been much better with visibly improved overall performance. The results of the above-mentioned ML and DL models are presented in Table 15.2. The CNN model performs better when fed with MFCC input rather than when fed with cleaned speech frames as input. All the simulations on Mozilla Common Voice dataset are carried out on an Intel i5 processor with 16GB RAM using Python 3.11. LibROSA package has been used to extract audio features. For training the CNN model, 80% of the dataset is used, whereas 10% is kept for validation and the rest 10% for testing.

15.5 CONCLUSIONS AND FUTURE SCOPE

The presented work on voice-based gender detection system is the outcome of a feature-oriented experimentation accompanied by ML and DL. Two potential features pitch and MFCC have been explored for the binary classification task. To establish the efficacy of the features as well as the ML models and CNN model, two publicly available datasets namely RAVDESS and Mozilla Common Voice have been considered. Since RAVDESS is an

Table 15.1 Performance on RAVDESS dataset

ML Classifier Model	*Accuracy (in %) with pitch as feature*	*Accuracy (in %) with MFCCs as feature*
SVM	80.3	99
Logistic Regression	79.1	99
DT	77.9	91.1
Ensemble (Bagged Tree)	86.5	95.8

Table 15.2 Performance on Mozilla Common Voice dataset with MFCC features

ML & DL Classifier	*Accuracy (in %)*	*Precision (in %)*	*Recall (in%)*	*Average Prediction Time*	*F1 Score (in %)*
SVM(ML)	95.36	95.47	95.28	0.0011	95.37
Logistic Regression (ML)	88.26	88.77	87.73	0.0010	88.24
Naïve Bayes (ML)	83.77	90.21	75.92	0.0130	82.45
DT(ML)	84.24	84.48	84.06	0.0015	84.27
Random Forest (ML)	84.28	84.81	85.29	0.0010	85.05
CNN(DL)	96.57	97.73	95.57	0.0840	96.63

emotion dataset created by recording emotional dialogues of male and female actors, there is a probability of the model outcomes being gender-biased. In this respect, Mozilla Common Voice dataset is a more appropriate dataset for the purpose of gender detection.

From the experiment, it is evident that MFCC is a more powerful feature than pitch for the gender detection task. MFCC feature-based ML models, i.e., SVM and Logistic Regression performed extremely well with 99% accuracy on RAVDESS dataset and with 95.47% and 88.77% accuracy on Mozilla Common Voice dataset. However, the CNN model performed better with 96.57% on Common Voice dataset. CNN, due to its massive connectionist structure, possesses better learning and generalization ability and hence performs exceedingly well on large datasets.

The work can be extended in different directions.

- Fusion of potential features is the most likely approach where features from time domain and frequency domain can be combined in many ways [20].
- Ensemble of ML classifiers is another way to achieve optimum performance. Different ensembling techniques can be followed to obtain maximum detection accuracy [21].
- Pretrained popular DL networks may also be exploited on larger datasets to gain maximum performance [22].

REFERENCES

1. Singh N., Agrawal A., and Khan R.A. (2018). Voice Based Biometric: A Technology for Voice Based Authentication. *International Journal of Advanced Science, Engineering and Medicine*, 10(7), 1–6.
2. Alkhawaldeh, Rami S. (2019). DGR: Gender Recognition of Human Speech Using One-dimensional Conventional Neural Network. *Scientific Programming*, 2019(12), 1–12.

3. Benzeguiba M., Mori R. D., Deroo O., Dupon S., Erbes T., et al. (2007). Automatic Speech Recognition and Speech Variability: a Review. *Speech Communication*, 49(10–11), 763–786.
4. Jasuja L., Rasool A., and Hajela G. (2020). Voice Gender Recognizer Recognition of Gender from Voice using Deep Neural Networks. In *International Conference on Smart Electronics and Communication (ICOSEC)*, IEEE, 2020.
5. Shabbir M., Hussain A., and Khan M.M. (2023). Age and Gender Estimation Through Speech: A Comparison of Various Techniques. In *18th International Conference in Emerging Technologies (ICET)*, Peshawar, Pakistan, IEEE, 228–233.
6. Ajmera J. and Burkhardt F. (2008). *Age and Gender Classification Using Modulation Cepstrum. Speaker Odyssey 2008: The Speaker and Language Recognition Workshop Stellenbosch*, South Africa, January 21–24, 2008.
7. Pahwa A., Aggarwal G.. (2016). Speech Feature Extraction for Gender Recognition. *International Journal of Image, Graphics and Signal Processing (IJIGSP)*, 8(9), 17–25.
8. Mavaddati S. (2024). Voice-based Age, Gender, and Language Recognition based on ResNet Deep Model and Transfer Learning in Spectro-temporal Domain. *Neurocomputing, Elsevier Science Publishers*. 580(1), 127429.
9. Benzeghiba M., De Mori R., Deroo O., Dupont S., Erbes T., et al. (2007). Automatic Speech Recognition and Speech Variability: A Review. *Speech communication*, Elsevier, 49(10–11):763–786.
10. Alnuaim A. A., et al. (2022). Speaker Gender Recognition Based on Deep Neural Networks and ResNet50.*Journal Wireless Communications and Mobile Computing, Wiley*, 2022(1) 1–13.
11. Nassif A. B., Shahin I., Attili I., Azzeh M. and Shaalan K. (2019). Speech Recognition Using Deep Neural Networks: A Systematic Review. *Journal, IEEE Acess*, 7, 19143–19165.
12. Yang K., Huang Z., Wang X. and Wang F. (2019).An SNR Estimation Technique Based on Deep Learning. International Journal, *Electronics*, 8(10), 1139.
13. Babu P.A., Nagaraju V.S. and Vallabhuni R.R. (2021). Speech Emotion Recognition System with LIbrosa. In *10th International Conference on Communication Systems and Network Technologies (CSNT)*, Bhopal, India, IEEE, 421–424.
14. Aurangzeb K., et al. (2021). Aspect Based Multi-Labeling using SVM Based Ensembler. *IEEE Access*, 9, 26026–26040.
15. Livingston A., Seow M. J., and Asari V. K. (2007). A Binary Decision Tree Based Real-Time Emotion Detection System. *Advances in Visual Computing, Springer, LNCS*, 4841, 432–441.
16. Jacob A. (2017). Modelling Speech Emotion Recognition Using Logistic Regression and Decision Trees. *International Journal of Speech Technology*, 20, 897–905.
17. Khan A. and Roy U. K. (2017). Emotion Recognition Using Prosodic and Spectral Features of Speech and Naïve Byes Classifier. In *International Conference on Wireless Communications, Signal Processing and Networking (WiSPNET)*, Chennai, IEEE, 1017–1021.
18. Yan S., Liang Y. E., Han S., Han T., Li Y., and Alasaarela E. (2020). Speech Interactive Emotion Recognition System Based on Random Forest. In

International Wireless Communications and Mobile Computing (IWCMC), Limassol, Cyprus, IEEE, 1458–1462.

19. Singh A., Srivastava K. K., Murugan H. (2020). Speech Emotion Recognition using CNN *in International Journal of Psychosocial Rehabilitation*, 24(8), 2408–2416.
20. McFee B., Raffel C., Liang D., Ellis D.P., McVicar M., Battenberg E. and Nieto O. (2015). Librosa: Audio and Music Signal Analysis in Python. In *Proceedings of the 14th Python in science Conference*, Vol. 8, 18–25.
21. Paul B., Bera S., Dey T., et al. (2024). Machine Learning Approach of Speech Emotions Recognition Using Feature Fusion Technique, *Multimedia Tools and Applications* 83, 8663–8688.
22. Alharbi Y. (2024). Effective Ensembling Classification Strategy for Voice and Emotion Recognition. *International Journal of System Assurance Engineering and Management*, Springer, 15, 334–345.

Chapter 16

Optimal dynamic contraflow with intermediate storage by anti-parallel path decomposition

Durga Prasad Khanal and Tanka Nath Dhamala

16.1 INTRODUCTION

The network flow problem in which shifting of the maximum amount of flow from the origin (source) node to the destination (sink) node without violating the capacity constraints on the arcs is known as maximum flow problem [11, 12]. This problem concerns with shifting the flow only at sink node and it does not permit the holding of the flow at intermediate nodes, that is, every intermediate node must satisfy the flow conservation constraint. They also have introduced the temporally repeated flow (TRF) and time expanded network to solve the maximum dynamic flow (MDF) problem. Gradual improvements on the solving techniques can be found in blocking flow algorithm [8], shortest augmenting path algorithm [9], push-relabel algorithm [14], and many more. The detailed illustrations of this problem and the solution strategies can be found in the survey papers of [2, 5, 21], book of [1], book series of [22] and their references.

In most of the literature, the network flow models allow the flow conservation constraints at the intermediate nodes, that is, inflow must be equal to outflow except at the origin and destination nodes. A new trend of research, known as flow with intermediate storage, is introduced by Pyakurel and Dempe [27] where flow out from the source is allowed to store at appropriate intermediate nodes, if it cannot reach the destination. The reason behind the storage of flow can be either the insufficient arc capacity to move forward or insufficient time to reach the next node from the current node. They introduced the maximum static and maximum dynamic flow problems and presented polynomial time solution strategies to solve the problems using lexicographically maximum flow approach. They also solved the universal maximum flow problem with intermediate storage in two-terminal series-parallel networks in [28]. Khanal et al. [16] extended the results in multi-commodity network flow to solve the maximum flow problem. Later on, Gerasimonko [13] solved the maximum dynamic flow model for hesitant fuzzy evacuation by allowing the storage of flow at intermediate nodes. Pyakurel et al. [26] extended the concept of flow with intermediate storage in an abstract network topology to solve the maximum static, lexicographic

DOI: 10.1201/9781003589716-16

maximum static, and maximum dynamic flow problems and presented algorithms to solve the problems in polynomial time complexity. Recently, the generalized maximum static and dynamic flow problems with intermediate storage in lossy network are introduced by Dhamala et al. [6] and they presented efficient algorithms for their solutions.

A two-way network is a network having at least a pair of nodes connected with anti-parallel arcs, that is, oppositely directed parallel arcs. The solution strategy used to solve the problems in such a network is the contraflow configuration in which the direction of opposite arcs is reversed toward the destination node to increase the amount of flow and reduce the traversal time. The complexity of the problem becomes *NP*-hard due to the decision process on the back or forth direction of each arc to get the best solution. The proof of *NP*-completeness of the contraflow problem with bounded evacuation time can be found in [19] by transforming the problem to 3SAT problem. Rebennack et al. [30] presented a polynomial time algorithm to solve the maximum flow problem in two-terminal network by reverting the direction of arcs at time zero and keeping them fixed afterward. Different aspects of contraflow configurations with various mathematical models, heuristics, optimization, and simulation techniques can be found in [3, 4, 10, 19, 20, 24, 25]. By incorporating the flow with intermediate storage, the maximum dynamic contraflow (MDCF) and universal maximum contraflow problems and their polynomial solutions can be found in [28] and [27], respectively.

Instead of reversing all the anti-parallel arcs, Pyakurel et al. [29] introduced the concept of partial contraflow in which only necessary arc capacities are reversed. In this technique, the remaining arc capacities are saved for other emergency purposes like facility allocation and logistic supports. All of these contraflow configurations discussed so far are the symmetric contraflow, where the transit times in anti-parallel arcs are the same. To handle the contraflow problem with different transit times on the anti-parallel arcs, Nath et al. [23] introduced the concept of orientation-dependent contraflow technique and solved the maximum dynamic and quickest contraflow problems in polynomial time. Allowing the intermediate storage, maximum dynamic contraflow problem in a general network and the earliest arrival contraflow problem in the two-terminal series-parallel network with asymmetric transit times can be found in [17]. The asymmetric contraflow configuration by using anti-parallel path decomposition and its use on solving maximum dynamic flow problem is presented in [15].

The chapter is organized as follows. Basic notations used throughout the chapter are presented in Section 16.2. In Section 16.3, we describe the different contraflow techniques for symmetric and asymmetric networks. The maximum dynamic contraflow (MDCF) problem with intermediate storage and its solution by anti-parallel path decomposition is presented in Section 16.4. The chapter is concluded in Section 16.5.

16.2 BASIC NOTATION

Consider a network $\mathbb{G} = (N, A, u, s, t)$, where N with $|N| = n$ represents a set of n nodes and $A \subseteq N \times N$ with $|A| = m$ represents a set of m arcs. Here, $s \in N$ and $t \in N$ are the source (origin) and sink (destination) nodes and $I = N \setminus \{s,t\}$ represents the set of intermediate nodes. Each arc $a = (v, w) \in A$ with $head(a) = w$ and $tail(a) = v$ has a capacity function $u : A \to \mathbb{R}^+$ that limits the flow on arc. Similarly, the set of outgoing arcs from node v and incoming arcs to node v are denoted by $\delta^{out}(v)$and$\delta^{in}(v)$, respectively. In case of dynamic network, two additional parameters are to be considered, one transit time $\tau : A \to \mathbb{R}^+$ that measures the transmission time from v to w and another $T = \{0, 1, \ldots, T\}$ to represent time horizon T in discrete time settings. Similarly, for the flow with intermediate storage, we also have to consider the capacity function on the nodes defined by $\nu : N \to \mathbb{Z}_0^+$ so that the dynamic network becomes $\mathbb{G} = (N, A, u, \nu, s, t, \tau, T)$.

16.3 CONTRAFLOW CONFIGURATION

A two-way network topology consists of pair of oppositely directed arcs (anti-parallel arcs) between some pair of nodes. Contraflow configuration technique is a very well-known technique on solving the network flow problems in such a network, which increases the outbound capacity of arcs by the reversal of oppositely orientated (backward) arcs toward the destination. In this section, we describe the symmetric and asymmetric network where the transit times on corresponding anti-parallel arcs are the same and different, respectively.

16.3.1 Symmetric contraflow

For a two-way dynamic network $\mathbb{G} = (N, A, u, \nu, s, t, \tau, T)$, consider that the transit times in anti-parallel arcs $a = (v, w)$ and $\bar{a} = (w, v)$ are the same, that is, $\tau_a = \tau_{\bar{a}}$ for all $a, \bar{a} \in A$. Then it is termed as a symmetric network. Let u_a and $u_{\bar{a}}$ be the capacity on a and $\bar{a}$, respectively. To solve the network flow problem with symmetric transit times, an auxiliary network is constructed as follows.

16.3.1.1 Auxiliary network

For a two-way network $\mathbb{G}$, an auxiliary network is denoted by $\overline{\mathbb{G}} = (N, \bar{A}, \bar{u}, \nu, s, t, \bar{\tau}, T)$, where $\bar{A}$ contains undirected (or bi-directed) edges $\bar{A} = \{(v, w) : (v, w) \text{or} (w, v) \in A\}$. The capacity $\bar{u}$ of an arc in the auxiliary network is the sum of capacities of anti-parallel arcs a and $\bar{a}$, that is,

$\bar{u}_a = u_a + u_{\bar{a}}$, where $u_a = 0$ if $a \notin A$. Similarly, the transit time of arc in an auxiliary network $\bar{\tau}$ is obtained by

$$\bar{\tau}_a = \begin{cases} \tau_a \text{ if } a \in A \\ \tau_{\bar{a}} \text{ otherwise} \end{cases}$$

All other parameters of $\bar{\mathbb{G}}$ are the same as in $\mathbb{G}$ (see Figure 16.1).

16.3.2 Asymmetric contraflow

In a two-way network $\mathbb{G}$, the traversal time in anti-parallel arcs may not always be the same. Network with different transit times in anti-parallel arcs (i.e., $\tau_a \neq \tau_{\bar{a}}$) is known as the network of asymmetric transit times. We can realize the asymmetric transit times on the one-way roads between two places or the curved and inclined/declined roads in mountain regions.

16.3.2.1 Orientation-dependent contraflow

For the contraflow configuration of asymmetric network, Nath et al. [23] introduced the contraflow with orientation-dependent transit times, where the transit time of reversed arc is taken as per its orientation. In this contraflow technique, the auxiliary network $\bar{\mathbb{G}} = \left(N, \bar{A}, \bar{u}, v, s, t, \bar{\tau}, T\right)$ of $\mathbb{G}$ is obtained having transit time $\bar{\tau}$ as

$$\bar{\tau}_a = \begin{cases} \tau_a \ \textit{if} \text{ orientaion is along the arc } a \\ \tau_{\bar{a}} \ \textit{if} \text{ the orientation is along the arc } \bar{a} \text{ or } a \notin A \end{cases}$$

The other components of $\bar{\mathbb{G}}$ are as similar to the contraflow with symmetric transit times (see Figure 16.2).

16.3.2.2 Contraflow with anti-parallel path decomposition

Here we present a technique of contraflow configuration in the network with asymmetric transit times in anti-parallel arcs, as in [15], termed as

(a) Arcs in symmetric network $\mathbb{G}$ (b) Auxiliary arc in $\bar{\mathbb{G}}$

Figure 16.1 Auxiliary arc (b) of given arcs in (a) with capacities and symmetric transit times.

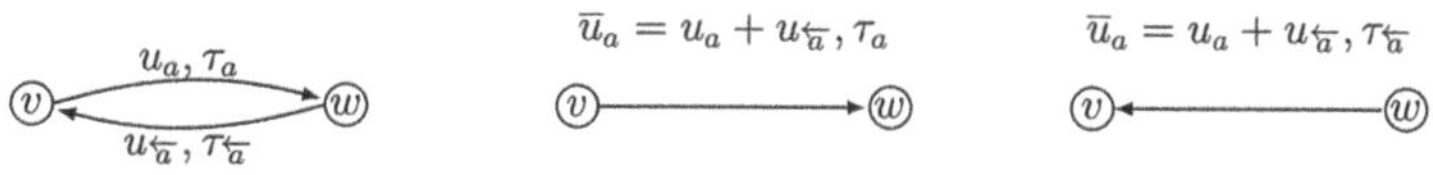

(a) Arcs in asymmetric network (b) Auxiliary arc if orientation is along a (c) Auxiliary arc if orientation is along $\overleftarrow{a}$

Figure 16.2 Orientation-dependent transit times in auxiliary arcs (b) and (c) of given arcs of an asymmetric network (a).

contraflow with anti-parallel path decomposition. For each given arc, we create an anti-parallel path of two arcs as follows. For an arc $a = (v, w)$ in the given network having u_a and τ_a as its capacity and transit time, we create an oppositely directed additional path $w - wv - v$ of two arcs $a_1 = (w, wv)$ and $a_2 = (wv, v)$ with $u_{a_1} = u_{a_2} = u_a$ and $\hat{\tau}_{a_1} = \alpha\tau_a, \hat{\tau}_{a_2} = (1-\alpha)\tau_a, 0 \le \alpha \le 1$, where wv is an artificial node along the path from w to v (see Figure 16.3). The network so obtained is termed as extended network $\hat{\mathbb{G}}$.

For a given network $\mathbb{G}$ with n nodes and m arcs, an extended network $\hat{\mathbb{G}}$ contains $m+n$ nodes and $3m$ arcs. The contraflow configuration in this technique depends on the extended network instead of auxiliary network which helps to solve the contraflow problem with unequal transit times on anti-parallel arcs. In the previous subsection, we have seen that the auxiliary network with orientation-dependent transit times modifies the properties of the arcs as per the change in directions. But this may not always be possible because the transit time for a road remains the same whatever be the direction of road, e.g. the one-way roads connecting two cities with parallel paths of different lengths.

An extended network $\hat{\mathbb{G}}$ is created in such a way that each arc $a \in A$ generates an anti-parallel path of two arcs without removing the original arc (see in Figure 16.3 having the same entities of its original arc). The extended network structure is represented as $\hat{\mathbb{G}} = (\hat{N}, \hat{A}, \hat{u}, s, t, \hat{\tau}, T)$ where $\hat{N} = N \cup \{v' : v' = vw \text{ or } wv\}, \hat{A} = A \cup \{a_1\} \cup \{a_2\}, \text{and } \hat{\tau}_a = \{\tau_a\} \cup \{\tau_{a_1}\} \cup \{\tau_{a_2}\}$. To this point, our main concern is to solve the maximum dynamic contraflow problem with intermediate storage by using anti-parallel path decomposition, where flow is not only shifted to the sink but also hold at prioritized intermediate nodes.

16.4 MAXIMUM DYNAMIC CONTRAFLOW (MDCF) WITH INTERMEDIATE STORAGE

In this section, we introduce the MDCF problem with intermediate storage, its flow model, and solution strategy by using anti-parallel path decomposition.

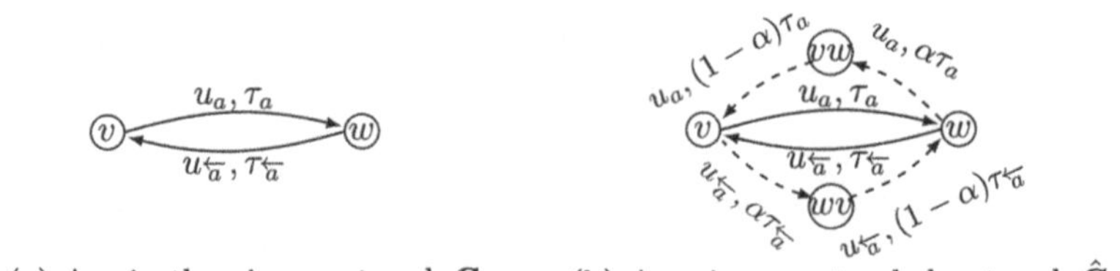

Figure 16.3 Anti-parallel path decomposition (b) of given arcs in (a) with capacities and transit times.

Problem 16.1

For a given two-way asymmetric network $\mathbb{G}$, the MDCF problem with intermediate storage by using anti-parallel path decomposition concerns with maximization of the flow at sink and prioritized intermediate nodes in an extended network $\hat{\mathbb{G}}$ within the given time horizon T.

16.4.1 Flow model

For the extended network $\hat{\mathbb{G}} = \left(\hat{N}, \hat{A}, \hat{u}, s, t, \hat{\tau}, T\right)$ of given network $\mathbb{G}$, let $\Phi : \hat{A} \times T \to \mathbb{R}^+$ be a dynamic $s-t$ flow function and $\psi : N \times T \to \mathbb{R}^+$ be the storage function. The linear programming contraflow model with intermediate storage defined on $\hat{\mathbb{G}}$ is.

$$\max |\Phi| \tag{16.1}$$

such that,

$$\sum_{a \in \delta_s^{out}} \sum_{\theta=0}^{T} \Phi_a(\theta) = |\Phi| = \sum_{a \in \delta_t^{in}} \sum_{\theta=\hat{\tau}_a}^{T} \Phi_a(\theta - \hat{\tau}_a) + \sum_{\upsilon \in N \setminus \{s,t\}} \psi_\upsilon(T) \tag{16.2}$$

$$\sum_{a \in \delta^{in}(\upsilon)} \sum_{\beta=\hat{\tau}_a}^{\theta} \Phi_a(\beta - \hat{\tau}_a) - \sum_{a \in \delta^{out}(\upsilon)} \sum_{\beta=0}^{\theta} \Phi_a(\beta) = \psi_\upsilon(\theta), \quad \forall \upsilon \in N \setminus \{s,t\}, \theta \in T \tag{16.3}$$

$$\sum_{a \in \delta^{in}(\upsilon')} \sum_{\beta=\hat{\tau}_a}^{\theta} \Phi_a(\beta - \hat{\tau}_a) - \sum_{a \in \delta^{out}(\upsilon')} \sum_{\beta=0}^{\theta} \Phi_a(\beta) = 0, \quad \forall \upsilon' = \upsilon w \text{ or } w\upsilon \in \hat{N} \setminus N, \theta \in T \tag{16.4}$$

$$0 \le \Phi_a(\theta) \le \hat{u}_a, \forall_a \in \hat{A}, \theta \in T \tag{16.5}$$

$$0 \le \psi_\upsilon(\theta) \quad \le \quad v_\upsilon, \quad \forall_\upsilon \in N \setminus \{s,t\}, \theta \in T \tag{16.6}$$

Here, Equation (16.1) is an objective function which aims to maximize the total flow. The total flow with intermediate storage, represented by Equation (16.2), is the total flow out from the source and is equal to the sum of flow at sink and intermediate nodes. Similarly, Equation (16.3) represents the excess flow over the time at intermediate nodes in N and Equation (16.4) is flow conservation constraint at each artificial node. Equations in (16.5) and (16.6) represent the arc capacity and node capacity constraints, respectively. As we aim to use the temporally repeated flow technique on the flow balancing paths, the node capacity must be T times the capacity of incoming arcs to the node in extended network, that is, $\nu_v \geq T\sum_{a\in\delta^{in}(v)} u_a, \forall a \in \hat{A}\cdot$

16.4.2 Solution strategy

The solution strategy begins with construction of an extended network $\hat{\mathbb{G}}$ of the given network $\mathbb{G}$. Then after, prioritization of nodes is fixed as follows: The first priority is given to the sink because it is considered as the most appropriate place. Priority of intermediate nodes is fixed as per the distance, that is, node farther in distance from the source is considered higher prioritized node. Thus, if $d(s,v_1) > d(s,v_2) >,\cdots,d(s,v_r)$ be the distance of intermediate nodes, then the set of prioritized nodes is $\{t,v_1,v_2,\cdots,v_r\}$. The excess flow and flow balancing paths are obtained in $\hat{\mathbb{G}}$ as follows.

16.4.2.1 Flow balancing paths with excess flow

Excess flow is the static flow at each node that cannot move to its successor node. The excess flow at the sink is the static flow at the sink calculated by minimum cost maximum flow algorithm. To calculate the excess flow of next prioritized intermediate node v_1, we first construct the residual network with reduced capacity. In this residual network, we find the static flow at v_1, which is excess flow at this node. This process is continued for second, third, and so on for each prioritized intermediate node unless there exists a flow carrying path to the node.

The next step is a very important step of finding flow balancing path $\hat{p}$ this, we take a network with excess flows at each node. Then, a shortest s-t path with positive flow is obtained from path decomposition of the static maximum flow. The flow on the path is balanced in such a way that the flow out from the source along the path is equal to the sum of the flow at the sink and excess flow at intermediate nodes. The network is then updated by reducing the arc capacities as well as node excess which is used by the first flow balancing path. This process is continued as long as there exists any s-t or s-v path with positive flow respecting the priority order of nodes and the process is terminated if no flow came out from the source. We present Algorithm 1 to obtain the flow balancing path with excess storage at nodes herein.

Algorithm 16.1

Flow balancing paths $\hat{p}$ with excess flow at nodes

Input: Given two-way dynamic network $\mathbb{G} = (N, A, u, s, t, \tau, T)$.

Output: Flow balancing path with excess flow.

1. Construct an extended network $\hat{\mathbb{G}} = (\hat{N}, \hat{A}, \hat{u}, s, t, \hat{\tau}, T)$ of given network $\mathbb{G}$.
2. Use the shortest path algorithm of [7] in $\hat{\mathbb{G}}$ to calculate the shortest distance of each intermediate node $N \setminus \{s,t\} \subset \hat{N}$.
3. Fix the first priority to the sink t and the priority order of intermediate nodes $v \in N \setminus \{s,t\} \subset \hat{N}$ with farther in distance from the source higher in priority order.
4. Obtain static flow at t to obtain the excess at t. Construct residual network and continue the process to find the static flow on each intermediate node $v \in N \setminus \{s,t\}$ with priority ordering in residual network to get excess flow at v.
5. Obtain the flow balancing paths $\hat{p}$ including excess at intermediate nodes as follows:
   (a) Take a shortest decomposed s-t path and balance the flow in such a way that flow out from the source along the path is equal to the sum of flow at the sink and excess flow at intermediate nodes.
   (b) The network is updated by reducing the flow on the arcs and the excess flow on the nodes.
   (c) Continue it as long as there exist s-t as well as s-v paths $\forall v \in N \setminus \{s,t\}$ with positive flow respecting the priority order.

16.4.2.2 Temporally repeated flow

Depending on the flow balancing paths $\hat{p}$ of previous subsection, we now present the formula that calculates the MDCF with intermediate storage on the asymmetric network.

Let f be a feasible static flow along the path $\hat{p} \in P$. The dynamic flow value at the sink node or the last node of the path is

$$\Phi_t = \sum_{\hat{p} \in P} \left(T - \tau(\hat{p}) + 1\right) f(\hat{p}) \tag{16.7}$$

Let $v, w \in N \setminus \{s\}$ are any two successive nodes in a flow balancing path $\hat{p}$ with $\hat{p}_{[s,v]} \subseteq \hat{p}_{[s,w]}$, then as introduced in [15, 18], the temporally repeated flow at v is

$$\psi_v(T) = \sum_{\hat{p}:\hat{P}_{[s,v]} \subseteq P} \left[\left(T - \tau\left(\hat{p}_{[s,w]}\right) + 1\right) \cdot \psi_v + \left(\tau\left(\hat{p}_{[s,w]}\right) - \tau\left(\hat{p}_{[s,v]}\right)\right) \cdot f\left(\hat{p}_{[s,v]}\right) \right] \tag{16.8}$$

where $\hat{p}_{[s,v]}$ is a flow balancing path from the source s to the node v, $\tau\left(\hat{p}_{[s,v]}\right)$ is the transit time of the path $\hat{p}_{[s,v]}$, ψ_v is the excess flow at the node v along $\hat{p}$, and $f\left(\hat{p}_{[s,v]}\right)$ is the static path flow from the source s to the node v.

The algorithmic framework to find the MDCF with Intermediate Storage by Anti-parallel Path Decomposition is presented as follows.

Algorithm 16.2

Temporally Repeated Algorithm of MDCF with Intermediate Storage by Anti-parallel Path Decomposition

Input: Two-way dynamic network $\mathbb{G} = (N, A, u, s, t, \tau, T)$.

Output: MDCF with Intermediate Storage by Anti-parallel Path Decomposition.

1. Construct an extended network $\hat{\mathbb{G}}$ of given network $\mathbb{G}$.
2. In $\hat{\mathbb{G}}$, fix the first priority to the sink t and the priority order of intermediate nodes $v \in N \setminus \{s,t\} \subseteq \hat{N}$ with farther in distance from the source higher in priority order.
3. Use Algorithm 16.1 to find the excess flow and flow balancing paths.
4. Obtain the MDF at sink and intermediate nodes by temporal repetition of flow using Equations (4.7) and (4.8).
5. MDF obtained in $\hat{\mathbb{G}}$ using Step 4 is the MDCF with intermediate storage by anti-parallel path decomposition.
6. Transform the solution to the original network by removing unused anti-parallel arcs and unused original arcs of the flow carrying anti-parallel arcs.

Theorem 16.1

Algorithm 16.2 provides an optimal solution to the MDCF problem with intermediate storage using anti-parallel path decomposition by respecting the priority order of nodes.

Proof. Construction of extended network $\hat{\mathbb{G}}$ is feasible because it does not change the attributes on its original arc and generates an optimal solution for contraflow problem. Calculation of excess flow in an extended network, obtained in Algorithm 16.1, is optimal because flow out from the source is either reached to the sink or the prioritized intermediate nodes. Similarly, flow balancing paths are obtained in such a way that, in each path, flow out from the source is equal to the sum of flow at sink and intermediate nodes. So, it provides optimal paths for static flow after reversal of arcs. To obtain the solution of MDCF problem with intermediate storage, we use Equations (16.7) and (16.8), which provides the solution in constant time respecting the priority order of nodes. Thus, the solution obtained in Algorithm 16.2 is an optimal solution to the MDCF problem with intermediate storage using anti-parallel path decomposition.

Example 16.1

As an example, consider a two-way network with asymmetric transit times on the arcs as shown in Figure 16.4(a), where the numbers on each arc represent the capacity and transit time. We assume that the storage capacity ν of each intermediate node be 150. The expanded network with anti-parallel path decomposition is presented in Figure 16.4(b) by taking $r = \frac{1}{2}$. As in Step 2 of Algorithm 16.2, first priority is given to sink and priority of intermediate nodes is fixed as per the distance from the source. As the shortest distance of v and w in $\hat{\mathbb{G}}$ are 1 and 2 units, respectively, the priority of nodes is $\{t, w, v\}$. Using Algorithm 16.1, the static excess flow is calculated as $\{13, 1, 3\}$ for $\{t, w, v\}$ which is presented via Table 16.1 and Figure 16.4(c).

The next step is to form the flow balancing paths $\hat{p}$. We choose one of the shortest paths, say $\hat{p}_1 = p_1$ from the Table 16.1, and balance the flow (see Figure 16.5). Then after the network is updated and next shortest path is taken for flow balancing. It is to be noted that this path may not be second

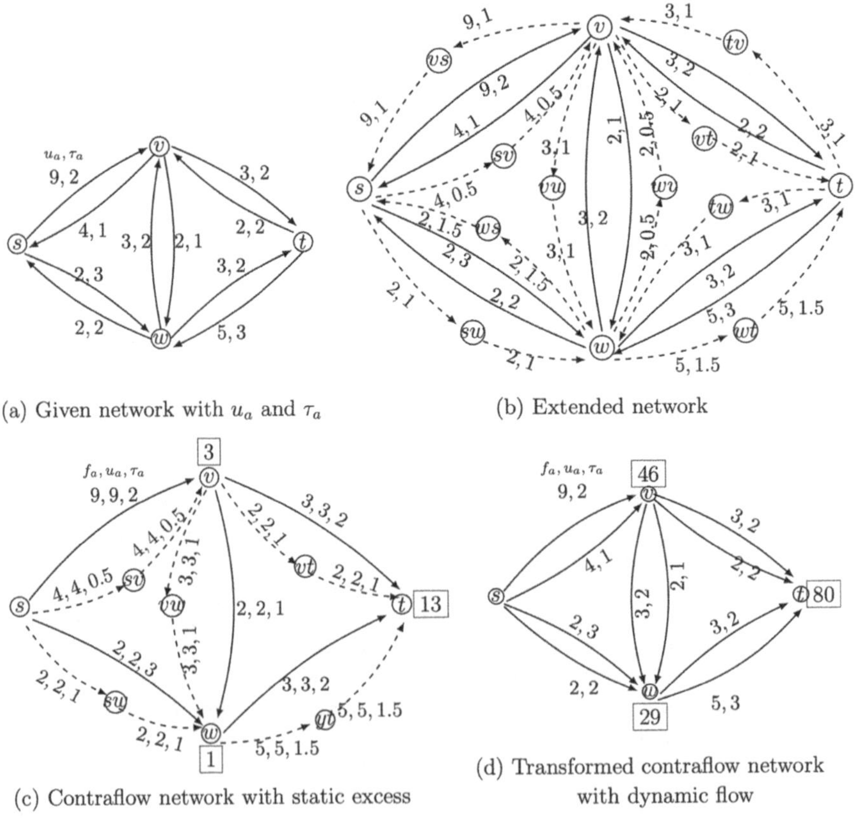

(a) Given network with u_a and τ_a

(b) Extended network

(c) Contraflow network with static excess

(d) Transformed contraflow network with dynamic flow

Figure 16.4 Contraflow with anti-parallel path decomposition.

Table 16.1 Static path flow with anti-parallel path decomposition

Path	*Transit time of path*	*Static flow* $(f(p))$
$p_1 : s - sv - v - t$	$\tau_{p_1} = 3$	3
$p_2 : s - sv - v - vt - t$	$\tau_{p_2} = 3$	1
$p_3 : s - v - vt - t$	$\tau_{p_3} = 4$	1
$p_4 : s - sw - w - t$	$\tau_{p_4} = 4$	2
$p_5 : s - w - t$	$\tau_{p_5} = 5$	1
$p_6 : s - w - wt - t$	$\tau_{p_6} = 6$	1
$p_7 : s - v - w - wt - t$	$\tau_{p_7} = 6$	2
$p_8 : s - v - vw - w - t$	$\tau_{p_8} = 7$	2
$p_9 : s - v - vw - w$	$\tau_{p_9} = 4$	1
$p_{10} : s - v$	$\tau_{p_{10}} = 2$	3

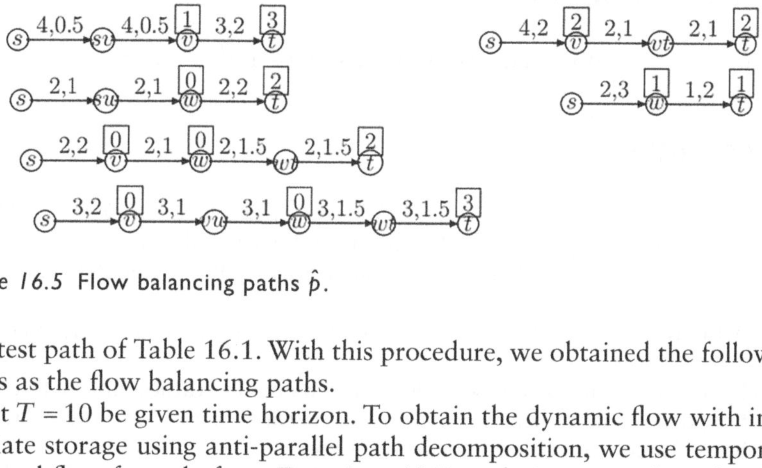

Figure 16.5 Flow balancing paths $\hat{p}$.

shortest path of Table 16.1. With this procedure, we obtained the following paths as the flow balancing paths.

Let $T = 10$ be given time horizon. To obtain the dynamic flow with intermediate storage using anti-parallel path decomposition, we use temporally repeated flow formula from Equations (4.7) and (4.8) to each node of the flow balancing paths. The total maximum dynamic contraflow within time horizon $T = 10$ are $|\Phi_t| = |\psi|_t = 80, |\psi|_w = 29$ and $|\psi|_v = 46$ units. Thus, total flow with intermediate storage is 155 units. The transformed network after contraflow configuration with dynamic flow value at nodes is presented in Figure 16.4(d).

16.5 CONCLUSION

In two-way network topology, contraflow is a very well-known technique to solve the network flow problems by reversing the direction of arcs toward the destination. Symmetric contraflow is well studied in literature, whereas asymmetric contraflow with orientation-dependent transit times, flow with intermediate storage, and asymmetric contraflow with anti-parallel path decomposition are actively growing areas of research. In this chapter, we have introduced the maximum dynamic contraflow problem by incorporating the intermediate storage of the flow and solved by anti-parallel decomposition. We have presented the flow model and its solution strategies. To the best of our knowledge, solution of the maximum dynamic contraflow problem with intermediate storage in asymmetric network by using anti-parallel path decomposition is for the first time.

CONFLICT OF INTEREST

The authors declare that there is no conflict of interest regarding the publication of this paper.

DATA AVAILABILITY

The authors have not used any additional data in this chapter.

REFERENCES

1. Ahuja, R.K., Magnanti, T.L. and Orlin, J.B. (1993). *Network Flows, Theory Algorithm and Applications*. Prentice Hall, Englewood Cliffs.
2. Aronson, J.E. (1989). A survey of dynamic network flows. *Annals of Operations Research*, 20, 1–66.
3. Arulselvan, A. (2009). *Network Model for Disaster Management*. PhD thesis, University of Florida USA.
4. Baumann, N., K'ohler, E.M. (2007), Approximating earliest arrival flows with flow-dependent transit times, *Discrete Applied Math*, 155, 161–171. DOI: https://doi.org/10.1016/j.dam.2006.04.030
5. Dhamala, T.N., Pyakurel, U. and Dempe, S. (2018), A critical survey on the network optimization algorithms for evacuation planning problems, *International Journal of Operations Research*, 15(3), 101–133.
6. Dhamala, T.N., Adhikari, M.C., Khanal, D.P., Pyakurel, U. (2024), Generalized maximum flow over time with intermediate storage, *Annals of Operations Research*. DOI: https://doi.org/10.1007/s10479-023-05773-w
7. Dijkstra, E. W. (1959), A note on two problems in connexon with graphs, *Numerische mathematik*, 1, 269–271. DOI: https://doi.org/10.1007/BF01386390
8. Dinic, E.A. (1970), *Algorithm for solution of a problem of maximum flow in a network with power estimation, Soviet Mathematics - Doklady. Doklady*, 11, 1277–1280. English translation by R.F. Rinehart.
9. Edmonds, J. and Karp, R. M. (1972), Theoretical improvements in algorithmic efficiency for network flow problems, *Journal of the ACM*, 19(2), 248–264.

10. Fleischer, L. and Tardos, E. (1998). Efficient Continuous-Time Dynamic Network Flow Algorithms. *Operations Research Letters*, 23, 71–80.
11. Ford, L.R. and Fulkerson D.R. (1958). A suggested computation for maximal multi-commodity network flows. *Rand Corporation*, 3, 27–58.
12. Ford, L.R. and Fulkerson D.R. (1962). *Flows in Networks*, Princeton University Press, Princeton, New jersey.
13. Gerasimenko, E., Kureichik, V., and Kuliev, E. (2022), Maximum Dynamic Flow Model for Hesitant Fuzzy Evacuation with Intermediate Storage at Nodes, In: Kahraman C., Cebi S., Cevik Onar S., Oztaysi B., Tolga A. C., Sari I. U. (eds) Intelligent and Fuzzy Techniques for Emerging Conditions and Digital Transformation. INFUS 2021. Lecture Notes in Networks and Systems, vol 307, (2022), Springer, Cham. DOI: https://doi.org/10.1007/978-3-030-85626-7_81
14. Goldberg, A.V. and Tarjan, R.E. (1988), A new approach to the maximum-flow problem, *Journal of the Association for Computing Machinery*, 35(4), 921–940. DOI: https://doi.org/10.1145/48014.61051
15. Khanal, D.P. (2023), Multi-commodity dynamic flow problems with intermediate storage and varying transit times, PhD Thesis, Institute of Science & Technology, Tribhuvan University, Nepal.
16. Khanal, D.P., Pyakurel, U., and Dhamala, T.N. (2021), Maximum multi-commodity flow with intermediate storage, *Mathematical Problems in Engineering*. DOI: https://doi.org/10.1155/2021/5063207
17. Khanal, D.P., Pyakurel, U., and Dempe, S. (2021), Dynamic contraflow with orientation dependent transit times allowing intermediate storage, *The Nepali Mathematical Sciences Report*, 38(2), 1–12. DOI: http://doi.org/10.3126/nmsr.v38i2.42700
18. Khanal, D.P., Pyakurel, U., and Dempe, S. (2022), *Temporally Repeated Maximum Dynamic Flow with Intermediate Storage, Preprint 2022-01, Fakultät für Mathematik und Informatik, TU Bergakademie Freiberg*, Germany.
19. Kim, S. and Shekhar, S. (2005), Contraflow network reconfiguration for evacuation planning: a summary of results, *In: Proceedings of 13th ACM Symposium on Advances in Geographic Information SystemsGIS*, vol. 05, 250–259.
20. Kim, S., Shekhar, S. and Min, M. (2008). Contraflow transportation network reconfiguration for evacuation route planning. *IEEE transactions on Knowledge and Data Engineering*, 20, 1115–1129.
21. Kotnyek, B. (2003). *An annotated overview of dynamic network flows.* Rapport de recherche 4936, INRIA Sophia Antipolis.
22. Kotsireas, I.S., Nagurney, A. and Pardalos, P.M. (2018). Dynamics of disasters-algorithmic approaches and applications, *Springer Optimization and Its Applications*, Springer Publication.
23. Nath, H.N., Pyakurel, U. and Dhamala, T.N. (2021). Network reconfiguration with orientation-dependent transit times. *International Journal of Mathematics and Mathematical Sciences*, Article ID 6613622, 11 pages, https://doi.org/10.1155/2021/6613622
24. Pyakurel, U. and Dhamala, T.N. (2016). Continuous time dynamic contraflow models and algorithms. *Advances in Operations Research – Hindawi* 2016, 1–7.
25. Pyakurel, U. Dhamala, T.N. and Dempe, S. (2017). Efficient continuous contraflow algorithms for evacuation planning problems. *Annals of Operations Research (ANOR)*, 254, 335–364.

26. Pyakurel, U., Khanal, D.P. and Dhamala T.N., Abstract network flow with intermediate storage for evacuation planning. *European Journal of Operational Research* 305 (3), 1178–1193, 2023. DOI: https://doi.org/10.1016/j.ejor.2022.06.054
27. Pyakurel, U., Dempe, S. (2020), Network flow with intermediate storage: models and algorithms, *SN Operations Research Forum*, 1, (2020), 1–23. DOI: https://doi.org/10.1007/s43069-020-00033-0
28. Pyakurel, U., Dempe, S. (2019), Universal Maximum Flow with Intermediate Storage for Evacuation Planning, In I. S. Kotsireas, A. Nagurney, P. M. Pardalos, and A. Tsokas A. (Eds.), Dynamics of Disasters, *Springer Optimization and Its Applications*, vol.169 (2021). Springer, Cham. DOI: https://doi.org/10.1007/978-3-030-64973-9\14
29. Pyakurel, U., Nath, H.N., Dempe, S. and Dhamala, T.N., (2019). Efficient dynamic flow algorithms for evacuation planning problems with partial lane reversal. *Mathematics*, 7, 1–29.
30. Rebennack, S., Arulselvan, A., Elefteriadou, L. and Pardalos, P.M. (2010). Complexity analysis for maximum flow problems with arc reversals. *Journal of Combinatorial Optimization*, 19, 200–216.

Chapter 17

Heart disease prognosis using machine learning classifier with hyperparameter optimization

Manas Kumar Swain, Lambodar Jena, Narendra Kumar Kamila, Soumen Nayak, and Swadhin Kumar Barisal

17.1 INTRODUCTION

A group of illness known as cardiovascular disease (CVD) affect the heart and blood vessels. The World Health Organization estimates that 1/5 of these global deaths occur in India. Age-standard cardiovascular death is reported by the Global Burden of Disease study. Cardiovascular disease is the leading cause of death worldwide, killing about 18 million people annually, with India having a rate of 272 per 100,000 people, higher than the global average of 235. This is roughly 30% of all fatalities worldwide. Heart attacks and strokes are responsible for more than 80% of deaths from cardiovascular disease, and 70% of those deaths happen before the age of 70. Cardiovascular disease is a major cause of heart failure, and this dataset contains 11 traits that can be used to predict the development of heart disease.

The term "CVD" refers to a number of conditions affecting the heart and circulatory system that present different pathologies like stroke, heart failure, or coronary disease. Heart failure is a pathophysiological condition in which the heart is unable to expel blood (unable to perform adequately). Heart rate (HR) can be classified according to the heart's ability to expel blood from the ventricles. Ejection Fraction (EF) is the percentage of the volume of blood expelled from the ventricle during systole and is determined by dividing the stroke volume by the end-diastolic volume. The ventricles typically constrict during systole and release 60 to 80% of the blood they hold by the end of diastole. Systolic or diastolic dysfunction may be the cause of heart failure.

By identifying trends in the data it receives from the past and basing its predictions on those trends, computers can learn on their own through a process known as machine learning. This is the best method for identifying diseases, because it can tell whether a patient has a CVD or not by looking at a number of crucial metrics. Several machine learning algorithms have been applied till now to diagnose different kinds of diseases. In this chapter, the LightGBM Algorithm has been implemented for the detection of the risks of heart diseases.

DOI: 10.1201/9781003589716-17

17.2 LITERATURE SURVEY

A brief discussion is presented on the research and implementation carried out by different authors around the globe to predict heart failure using machine learning algorithms more efficiently.

In Budholiya et al. (2022), a proposal was made to diagnose heart disease. The proposed work uses One-Hot coding to optimize eXtreme Gradient BOOSTing (XGBoost) and five different assessment metrics: precision, sensitivity, specificity, F1 score, and Area Under the ROC curve (AUC) of the Receiver Operating Characteristic (ROC) charts. It also uses Bayesian optimization as a hyperparameter optimization technique which has proven to be a very accurate method for obtaining the best hyperparameters.

Kaushik et al. (2021) were able to accurately assess and stratify the risk of heart failure of the patients using the XGBoost algorithm. It uses the GridSearchCv to select the parameters with maximum accuracy. It also used SHapley Additive exPlanations (SHAP)values that explicitly explained key features that influenced the model's result. It helped doctors understand why people suffer from heart failure.

The analysis was carried out on 299 heart failure patients in 2015. 299 patients with heart failure underwent the analysis in 2015. Chicco et al. (2020) also used conventional biostatistical tests to classify alternative traits, and they compared the outcomes to those of the algorithm.

The survival predictive performance using gender-specific informative risk factors for patients with left ventricular systolic dysfunction was studied by Zahid et al. (2019). The authors concluded that the predictive model for women differs significantly from that for men. Platelet count, while for men smoking, diabetes, and anemia with zero regression coefficients are not meaningful.

In order to maximize the effectiveness of the random forest classifier and the XGBoost classifier model, the documentation relies on hyperparameter-optimization using randomized search, grid search, and genetic programming (Panjaitan et al., 2023). These two models' performance was compared to that of prior research.

The analysis of Mathers et al.'s (2006) study was based primarily on optimistic express assumptions and represented one of three hard and fast visions of the future for population fitness. Based on the data from 1990, the predictions were made. Even though they were very out of date, they anticipated the spread of HIV/AIDS. Making public-fitness officers plan future regulations to reduce the burden of disease and deaths globally was the authors' main objective.

Kaur et al. (2022) have created a hybrid genetic-fuzzy coronary heart disorder analysis tool. The suggested version, known as the GAFL (Genetic Algorithm Fuzzy Logic), produces an accurate prediction of coronary heart disease. Because it was simple to construct, the authors' main objective was to provide resources to doctors working in hospitals and other clinical settings.

In this article (Khourdifi et al., 2019), the authors developed a PSO-optimized multi-layered perceptual artificial neural network and applied the Fast Correlation-Based Feature Selection (FCBF) method to remove redundant features and enhance the classification of heart diseases. The proposed model's mixed strategy is applied to the heart disease dataset. The model produced an accuracy rate of 99%.

The authors of this article (Babaoğlu et al., 2010) demonstrate how effectively artificial intelligence techniques are used in medical diagnostic tools to improve diagnostic accuracy and give doctors access to more information about. About 500 patients with 23 characteristics made up the dataset and it provides output with an accuracy rate of 79%.

The SMOTE oversampling algorithm, which was created to address the imbalance in datasets, was described in the article (Chawla et al., 2002). In order to improve the performance of the classifier in the ROC space, this method performs oversampling in the minority class and subsampling in the majority class.

17.3 OBJECTIVE

In this section, the chapter's goals are clearly stated. The chapter has a number of goals.

i. The main objective of the chapter is to demonstrate the importance of data pre-processing in raising a machine learning model's general performance.
ii. This chapter aims to predict whether or not someone will suffer from heart disease or not more accurately than the already present methods. Based on numerous features like age, sex, chest pain, resting blood pressure, cholesterol, etc., the results are predicted.
iii. This chapter additionally exploits the overall performance of different algorithms and compares them with the proposed version of this chapter, which includes linear regression, K-Nearest Neighbors (KNN), random forests, support vector machines (SVM), and XGBoost.
iv. Finally, SHAP values are used to interpret the received results in order to assess the effect of every feature on the prediction.

17.4 PROPOSED MODEL

LightGBM is the suggested model for estimating the likelihood that a patient will suffer from heart diseases. An open-source library called Light GBM offers a more effective and precise implementation of gradient boosting algorithms. It is an entire, distributed gradient boosting framework with high overall performance that is entirely based on decision tree algorithms.

These are employed in a variety of system learning tasks, including ranking and classification (Omotehinwa et al., 2023; Navazi et al., 2023; Ansarullah et al., 2022).

The hyperparameters of LightGBM are tuned in this chapter, and it is discovered that the tuning resulted in a higher accuracy rating. Positive overall performance metrics have been used to analyze the overall performance of the proposed version. The maximum accuracy rate for the results is 94.69%. Furthermore, the proposed version indicates a most F-score value of 94.75% and a precision of 93.74%. The outcomes of this technique are compared with numerous present algorithms which include Linear Regression, KNN, Random Forests, and XGBoost, and it is found that the accuracy of LightGBM with tuned hyperparameter outperformed them in terms of accuracy, precision, and F-score values. As a result, the proposed model can more accurately predict the risk of patients who could develop heart diseases.

LightGBM models require extra knowledge and tuning than Random Forest and Support Vector Machines techniques. It is nearly seven times quicker than XGBoost and is a far better method when the datasets are huge. This library is mainly focused on computational speed and model performance; as such there are few frills. The implementation of the set of rules turned into the performance of computing time and memory resources.

In this chapter, the layout goal was to make quality use of assets to teach the model. The gradient boosting decision tree algorithm is implemented by LightGBM. LightGBM basically operates on the tabular data, with the rows representing observations, one column representing the goal variable or label, while the final columns representing features.

We undergo cycles that again and again construct new models and integrate them into an ensemble model. We begin the cycle with the means of calculating the error for every observation of the dataset. Then we construct a new model to predict the results. The predictions are uploaded from the model to the "ensemble of models." All the predictions are now uploaded in the preceding order to make the prediction. Those predictions can be used to calculate new errors, construct a subsequent model, and then upload them to the ensemble. A few base predictions are needed to begin the cycle. Even if its predictions are very inaccurate, the next addition to the ensemble will deal with the errors.

LightGBM has some parameters which could affect our model's accuracy and training speed. The parameters altered on this model are learning rate, min child weight, gamma, subsample, col sample by tree, and max depth. These parameters were altered using the iterative randomized search which tried all the permutations and combinations and selected the values which resulted in the maximum accuracy.

This work is very much focused on data pre-processing as it plays a very vital role in developing an efficient model. The null and categorical values were mainly taken care of in this process. For the starters, the null values were filled with the median of the other values present in that column. Then, the features including gender, ever married, and residence type were converted into integer values from string values using the map function. Due to the presence of class imbalance, SMOTE was used to create an over-sample of the minority class. This drastically increases the accuracy of the model. The dataset is now divided into parts–Training data and Testing data using a sci-kit learn tool, train_test_split. The dataset is split in this kind of manner that the training data and testing data include 80% and 20% of the dataset, respectively. The model is now trained using the LightGBM classifier. To obtain the maximum accuracy, a set of grid parameters are considered for the LightGBM. The iterative randomized search, which applied all permutations and combinations that resulted in the maximum accuracy, was used to extract the best value combinations from the grid. The model is now being trained using these obtained values. The name of this technique is hyperparameter tuning. The results are then obtained by comparing the trained model to the testing data. Figure 17.1 shows a diagram that illustrates this process.

17.5 RESULTS AND ANALYSIS

We have discussed the outcomes of using the LightGBM classifier in this section. The outcomes were contrasted against those of other techniques like Logistic Regression, KNN, SVM, and XGBoost. The model was trained and tested on a dataset with 5000 (approx.) cases. On testing the model against

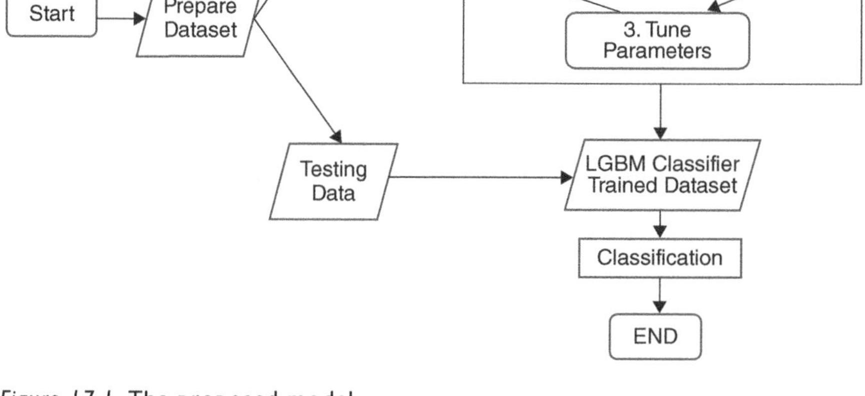

Figure 17.1 The proposed model.

the testing data, the LightGBM classifier-trained model resulted in an accuracy of 94.95%. The recall, precision, and F-score were 95.88%, 94.13%, and 94.99%, respectively. In Figure 17.2, the performance is represented under the ROC curve.

The results were further compared with the results obtained from other methods. It was found that the LightGBM outperformed other methods in terms of accuracy including Logistic Regression (LR), KNN, Random Forests (RF), and XGBoost which were found to be 79.25%, 82.61%, 86.27%, and 90.91%, respectively. The comparison is shown in Figure 17.3.

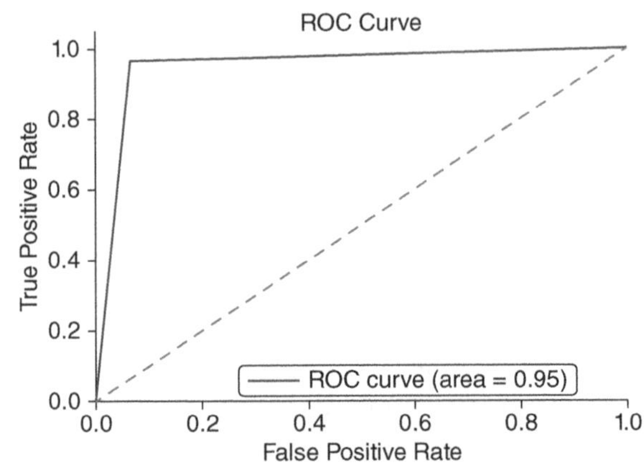

Figure 17.2 ROC curve representing TPR vs FPR.

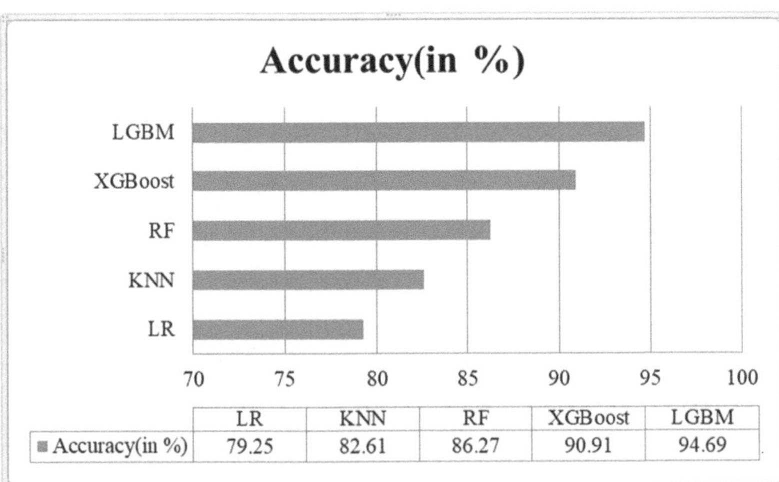

Figure 17.3 Comparative analysis of the accuracy of LightGBM with other algorithms.

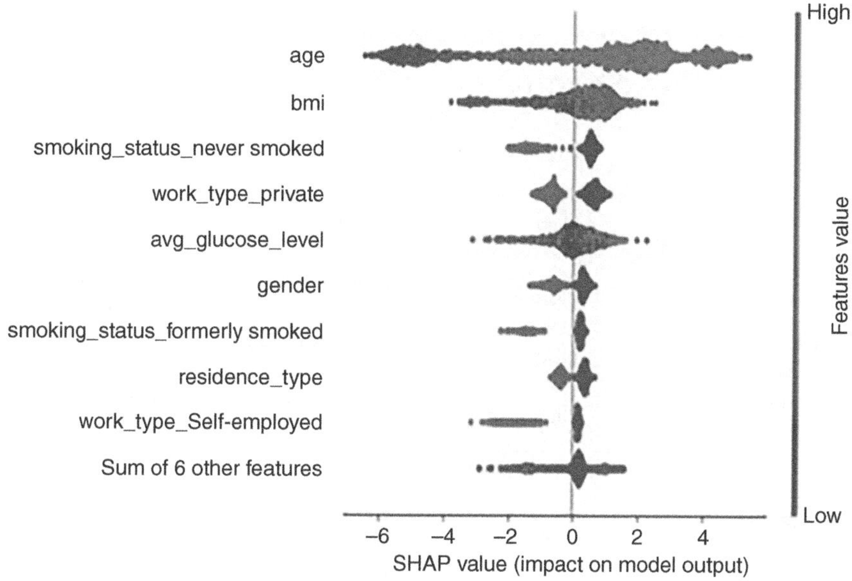

Figure 17.4 Impacts of each feature on the output model.

We also chose the functions that had an impact on the outcomes of using SHAP values. In The plot in Figure 17.4 shows all kinds of features through the sum of SHAP value magnitudes over all samples. It is observed that the patients who suffered from heart diseases showed features like higher age, had smoking habits, average glucose levels, gender, and work type.

The X-axis of the plot represents the SHAP value, i.e., impact on model output. The SHAP value helps to explain how much each feature contributes to pushing the model's prediction higher or lower. The negative values mean a feature decreases the predicted outcome and the positive values mean a feature increases the predicted outcome.

The Y-axis of the plot represents the feature names. The features used in the model are ranked in order of their importance (top to bottom). The features that are more important appear higher.

Table 17.1 describes the impact of each feature on model output.

The contribution of every feature is explained in Figure 17.5. The figure demonstrates the contribution of the features in the output given by the model. Features that tend the prediction to class 1(i.e., Heart Disease) are represented in horizontal bars pointing toward right (forward direction) while the features tending the prediction to class 0 (i.e., no Heart Disease) are represented in horizontal bars pointing toward left (backward direction).

Table 17.1 The impact of each feature on model output

Feature	*Impact on Model Output (SHAP Value)*	*Effect of High Values*	*Effect of Low Values*
Age	High	Increases model output	Decreases model output
BMI	High	Mixed impact (positive & negative)	Mixed impact
Never Smoked	Moderate	Slightly decreases output	Slightly increases output
Work Type – Private	Moderate	Increases model output	Decreases model output
Average Glucose Level	High	Increases model output	Decreases model output
Gender	Moderate	Slight increase in output	Slight decrease in output
Formerly Smoked	Low	Slightly increases output	Slightly decreases output
Residence Type	Low	Slight increase in output	Slight decrease in output
Work Type - Self-Employed	Low	Slight increase in output	Slight decrease in output
Sum of 6 Other Features	Low	Mixed impact	Mixed impact

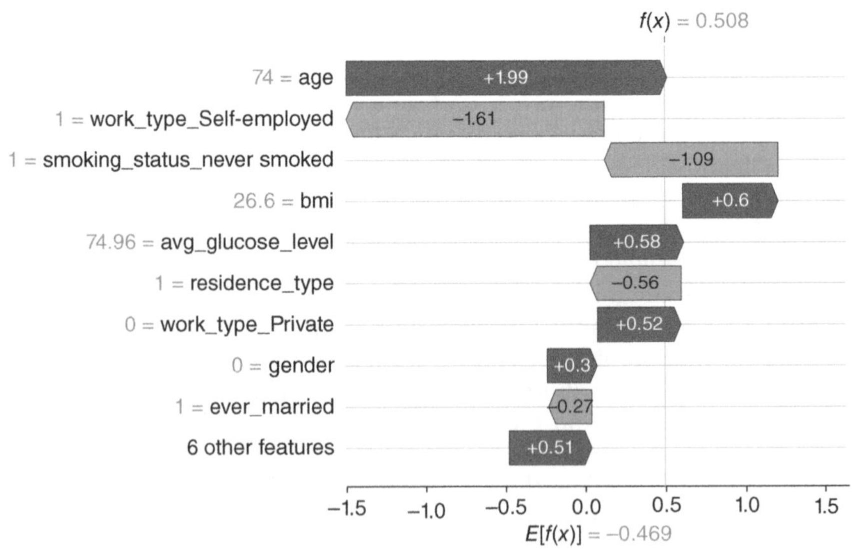

Figure 17.5 Contribution of each feature in the model results.

17.6 CONCLUSION

Heart disease is a leading cause of death each year. The goal of this chapter was to estimate the likelihood that patients will experience heart disease most frequently. We have concentrated on the imbalance in the dataset, a very prevalent issue, in order to improve the model's efficacy. It was found that using SMOTE significantly improved the model's accuracy. By tuning the hyperparameters of the LightGBM classifier, we obtained an accuracy of 94.69% (~95% approx.). The global and local SHAP values were calculated to explain how each feature affected the model as well as how each patient's symptoms contributed to the diagnosis for that patient. From the SHAP value plots, it can be concluded that: Older patients are more at risk of having heart disease. Patients with low BMI scores are less likely to suffer from heart diseases. Patients with smoking habits are at a higher risk to suffer from heart diseases. High average glucose level also increases the risk of heart diseases. Males are more likely to develop heart disease than females. Self-employed patients are less likely to suffer from heart diseases.

REFERENCES

Ansarullah, S.I., MohsinSaif, S., Abdul BasitAndrabi, S., Kumhar, S.H., Kirmani, M.M., and Kumar, D.P.An intelligent and reliable hyperparameter optimization machine learning model for early heart disease assessment using imperative risk attributes. *J Healthc Eng*. 2022 (2022 Apr 12), 9882288. doi: 10.1155/2022/9882288. PMID: 35449846; PMCID: PMC9018172

Chicco, D. and Jurman, G. Machine learning can predict survival of patients with heart failure from serum creatinine and ejection fraction alone. *BMC Med Inform DecisMak* 20 (2020), 16. 10.1186/s12911-020-1023-5

ChawlaNitesh, V., Bowyer, K.W., Hall, L. O., and Philip, K. W., SMOTE: Synthetic minority oversampling technique, *J Artif Intell Res* 16 (1) (2002), 321–357

Panjaitan, F., Nurmaini, S., and Partan, R. U. Hyperparameter Optimization in Machine Learning for Heart Disease Classification based on Heart Rate Variability Feature Analysis.*2023 International Conference on Informatics, Multimedia, Cyber and Informations System (ICIMCIS), Jakarta Selatan, Indonesia*, 2023, pp. 393–397, doi: 10.1109/ICIMCIS60089.2023.10349050

Navazi, Fatemeh, Yuan, Yufei, and Archer, Norm. "An examination of the hybrid meta-heuristic machine learning algorithms for early diagnosis of type II diabetes using big data feature selection". *Healthc Anal*, 4 (2023), 100227, ISSN2772-4425, 10.1016/j.health.2023.100227

Babaoğlu, Ismail, Fındık, Oğuz, and Mehmet, Bayrak. Effects of principle component analysis on assessment of coronary artery diseases using support vector machine. *Expert Syst Appl*, 37 (3) (2010), 2182–2185, ISSN 0957-4174, 10.1016/j.eswa.2009.07.055

Budholiya, Kartik, Shrivastava, Shailendra Kumar, and Sharma, Vivek. An optimized XGBoost based diagnostic system for effective prediction of heart disease.

J King Saud Univ ComputInf Sci, 34 (7) (Jul 2022), 4514–4523. 10.1016/j.jksuci.2020.10.013

Kaur, J. and Khehra, B.S. Fuzzy logic and hybrid based approaches for the risk of heart disease detection: State-of-the-art review. *J Inst Eng India Ser B* 103 (2022), 681–697. 10.1007/s40031-021-00644-z

Khourdifi, Youness and Bahaj, Mohamed. Heart disease prediction and classification using machine learning algorithms optimized by particle swarm optimization and ant colony optimization. *Int J Intell Eng Syst* (2019). 10.22266/IJIES2019.0228.24

Mathers, C.D. and Loncar, D. Projections of global mortality and burden of disease from 2002 to 2030. *PLoS Med* 3(11) (2006 Nov), e442. 10.1371/journal.pmed.0030442. PMID: 17132052; PMCID: PMC1664601.

Kaushik, S. and Birok, R. Heart Failure prediction using Xgboost algorithm and feature selection using feature permutation. *2021 Fourth International Conference on Electrical, Computer and Communication Technologies (ICECCT)*, Erode, India, 2021, pp. 1–6, doi: 10.1109/ICECCT52121.2021.9616626

Omotehinwa, Temidayo Oluwatosin, Oyewola, David Opeoluwa, and Dada, Emmanuel Gbenga. A light gradient-boosting machine algorithm with tree-structured Parzen Estimator for breast cancer diagnosis, Healthcare Analytics. *Healthcare Analytics* 4 (2023), 100218, ISSN 2772-4425, 10.1016/j.health.2023.100218

Zahid, F.M., Ramzan, S., Faisal, S., Hussain, I. Gender based survival prediction models for heart failure patients: A case study in Pakistan. *PLoS One* 14(2) (2019 Feb 19), e0210602. 10.1371/journal.pone.0210602. PMID: 30779736; PMCID: PMC6380566.

Index

Pages in *italics* refer to figures and pages in **bold** refer to tables.